1978

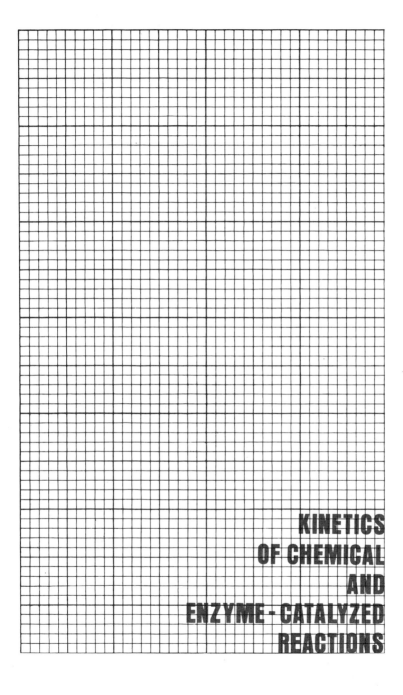

KINETICS
OF CHEMICAL
AND
ENZYME - CATALYZED
REACTIONS

KINETICS
OF CHEMICAL
AND
ENZYME - CATALYZED
REACTIONS

Department of Biological Chemistry
California College of Medicine
University of California, Irvine

New York
OXFORD UNIVERSITY PRESS
1977

PREFACE

The discipline of kinetics studies the rates of reactions for the ultimate purpose of understanding their mechanisms. Kinetics has important applications in the continuum which extends from chemistry to biochemistry; it is the primary tool used to deduce how simple chemical catalysts and complex enzymes accelerate reactions.

This slim volume is intended to serve several functions. It is meant as a means of teaching kinetics to advanced undergraduate and graduate students. It presupposes a background in organic chemistry and some knowledge of calculus. While a familiarity with biochemistry and chemistry would be useful, it is not essential. This book is also meant to be a primer for reading papers in the literature dealing with kinetics. In addition, I hope that it may serve as an introductory, practical handbook for the novice experimenter; in its format it should stand as a compact reference of common methods of graphical analysis. As such, it is intended as a first source, although not a last resort.

I wish to acknowledge the many people who have helped me in preparing the manuscript. I owe special thanks to Dr. Kenneth Ibsen whose many discussions and suggestions helped to shape the outline of this text. I also thank Dr. Larry Overman, Robert Goitein, and John Moe for their suggestions and critical review of the manuscript. Lastly, I must thank Mrs. Evelyn Grace for her patience and dedication in converting my marginally legible, handwritten rough draft into this final version.

D.P.

Laguna Beach, California
September 1976

CONTENTS

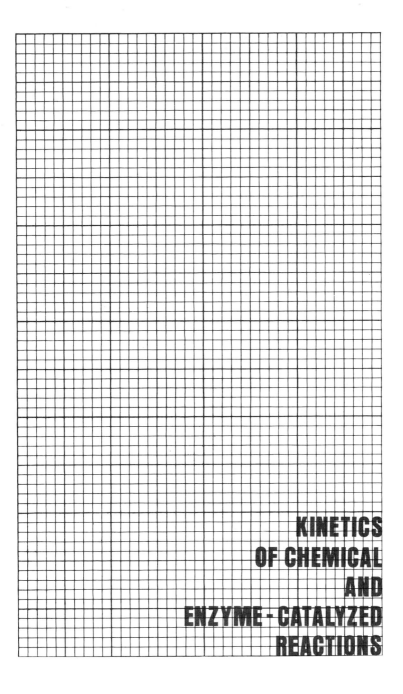

KINETICS
OF CHEMICAL
AND
ENZYME - CATALYZED
REACTIONS

ENZYMES AND CATALYSIS

Probably the first chemical reaction sequence used by man was the fermentation process which produces ethyl alcohol. Several millennia passed as our ancestors utilized this discovery before its cause was identified. In 1871 Pasteur demonstrated that living yeast cells were required for alcoholic fermentation, and in 1897 Buchner discovered that a cell-free yeast extract could also bring about fermentation. In the years which followed, numerous investigators have contributed to an understanding of alcoholic fermentation as a complex process by which yeast cells, through a series of chemical reactions, convert glucose to ethyl alcohol. The individual reactions of this process are brought about by enzymes —biological catalysts which are more efficient and specific than any devised by man.

Enzymes have been found to be responsible for nearly all chemical reactions that occur in living organisms. Many enzymes, such as the proteolytic enzymes of the digestive tract, are secreted and function outside the cell. However, most enzymes function as catalysts of distinct steps of highly organized metabolic pathways. Thus, while an enzyme and the reaction it catalyzes may be considered individually, it should be remembered that all enzymes serve as part of an organized (hence, the term *organic*) living system.

Enzymes, from the point of view of their chemical nature, are proteins. In 1926 Sumner was able to obtain the enzyme urease, which hydrolyzes urea to NH_3 and CO_2, as protein crystals. Initially this achievement was met with some skepticism based on the possibility that the enzyme was a trace impurity in the crystalline protein. However, in succeeding years this doubt faded as Northrup, Kunitz, and others succeeded in preparing additional enzymes as crystalline proteins. Our understanding of enzymes as proteins has grown rapidly since then. In 1963 Moore and Stein elucidated the amino acid sequence (the complete covalent structure) of ribonuclease, and in 1965 Phillips reported the three-dimensional structure of lysozyme. These firsts in determination of covalent and three-dimensional structure were rapidly followed by similar successes with other enzymes. The result of this work is that today numerous enzymes have achieved the status of extremely well-characterized proteins.

In 1894 Ostwald formulated the modern definition of a catalyst as a substance that increases the rate of a chemical reaction, but does not appear in the balanced chemical equation. This defini-

3

tion is valid for both chemical and enzyme catalysts, and with it as a starting point we may consider the dynamic properties of enzymes in greater detail.

Enzymes are effective in small amounts. The concentration of an enzyme required to bring about the complete conversion of reactant, generally termed *substrate,* to product need be only a minute fraction of the concentration of the reactant. The concentrations of enzymes within cells seldom exceed 10^{-5}M and are generally several orders of magnitude lower. However, substrate concentrations may be as high as 10^{-4} to 10^{-3}M.

Enzymes are unchanged by the reaction. This statement follows logically from the observation that enzymes function in trace quantities. To do so they must be regenerated, or "turn over," after conversion of one molecule of substrate to product. After regeneration, the enzyme may convert additional molecules of substrate to product, being regenerated after each conversion. Enzymes may turn over as few as 20 or as many as 3,000,000 molecules of substrate per molecule of enzyme per minute as is the case for catalase.

Enzymes do not affect the equilibria of reversible chemical reactions; they only accelerate the rates at which the equilibria may be reached. This property of enzymes is often expressed quantitatively in thermodynamic terms. (The applicable thermodynamics are treated superficially here. A more detailed discussion of this subject, especially the experimental determination of thermodynamic parameters from equilibrium and kinetic data, is given in Chapter 3.) The second law of thermodynamics states that any system not at equilibrium may spontaneously change toward a state of equilibrium. If a reaction proceeds spontaneously toward equilibrium and is opposed by a force tending to reverse it, the reaction may be made to do useful work, i.e., expend energy. The maximum useful work that can be obtained from a chemical reaction is termed the change in free energy of the reaction. This change in free energy is denoted ΔG. (In older texts free energy is often denoted as F. The newer convention is to use G in honor of Josiah Willard Gibbs, who pioneered the development of this concept.) Once a system has reached equilibrium, it is incapable of doing useful work; and therefore the available free energy is zero.

The variation of free energy during the course of a reaction explains in thermodynamic terms how an enzyme can accelerate the rate of a reaction. The variation of free energy during the course

4

of an uncatalyzed reaction, a chemically catalyzed reaction, and an enzyme-catalyzed reaction is shown graphically in the reaction coordinate diagram (Fig. 1-1). According to this diagram the free energy contents of reactant and product are set; their positions are fixed regardless of the nature of catalysis or absence of it. The amount of energy released upon reaction has already been defined as ΔG, and this quantity is always negative for a spontaneous reaction. While energy may be released upon reaction, the reaction

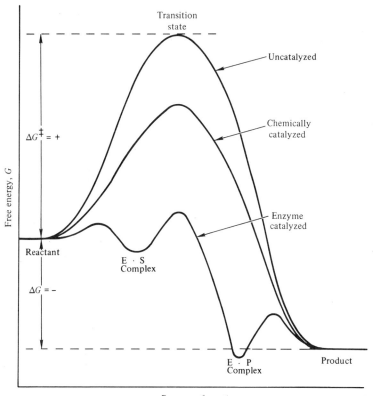

Fig. 1-1. Reaction coordinate diagram for idealized uncatalyzed, chemically catalyzed, and enzyme-catalyzed reactions. The positions of enzyme-substrate and enzyme–product complexes are given at E·S and E·P, respectively.

may not necessarily take place rapidly. A certain amount of additional energy may be required by the system before reaction can take place. This added energy is required to lift the reactant to the energy level of a *transition state,* or *activated complex,* which is a short-lived, unstable species with a structure intermediate between reactant and product. Chemists often initiate reactions by delivering energy in the form of heat to the reactants. The additional energy required to initiate reaction is the free energy of activation, $\Delta G^{\ddagger}$. The free energy of activation, $\Delta G^{\ddagger}$, can be calculated as an inverse function of the experimentally determined reaction rate (see Chapter 3). The slower the rate of reaction the higher $\Delta G^{\ddagger}$, and vice versa. The uncatalyzed reaction has a relatively large value for $\Delta G^{\ddagger}$. The chemically catalyzed reaction proceeds at a faster rate than the uncatalyzed reaction and has a smaller $\Delta G^{\ddagger}$ value. The enzyme-catalyzed reaction proceeds faster than either the uncatalyzed or chemically catalyzed reaction, and it has the smallest $\Delta G^{\ddagger}$ value.

Both chemical and enzyme catalysts function in lowering the free energy of activation, $\Delta G^{\ddagger}$, by changing the mechanism of the reaction. In other words, a catalyst alters the nature of the intermediates on the reaction pathway between reactant (or substrate) and product. As we will see in a later chapter, enzymes form a complex with substrate prior to reaction and produce a complex of enzyme and product. Both the enzyme-substrate complex and the enzyme-product complex have free energy levels lower than that of the transition state of the uncatalyzed reaction. These complexes are metastable and have local free energy minima. The positions of the intermediates formed in the idealized enzyme-catalyzed reaction are shown in the reaction coordinate diagram (Fig. 1-1). Thus, in an uncatalyzed reaction, the reactant must climb a mountain of free energy before it can descend into the valley of the product. In the chemical and enzyme-catalyzed reactions, the reactant takes a faster pathway across the low foothills.

Enzymes are usually specific in both the nature of the reaction they catalyze and the structures of their substrates. To illustrate this point, acid (hydrogen ion) may function as a chemical catalyst in a wide variety of reactions; it may catalyze the hydrolyses of esters, amides, and glycosides of a wide variety of structures. Lysozyme will hydrolyze only glycosides of a given structure, albeit at a much faster rate than acid could. In 1894 Emil Fisher noted this specificity and attributed it to a complementarity of structures

6

of substrate and enzyme; the substrate fits the surface of the enzyme as specifically as a key fits into a lock. This "lock and key" theory required a rigidity of enzyme and substrate which is no longer believed necessary. However, this theory has directed our understanding of the mechanisms of enzyme-catalyzed reactions. It led directly to the concept of the "active site" as that region on the surface of an enzyme where catalysis is effected.

Enzymes generally function under moderate conditions. Chemical catalysts often catalyze reactions under extreme conditions: elevated temperature, elevated pressure, high concentrations of acid or base. In contrast, enzymes generally function at temperatures ranging from approximately 10°C to 50°C and pH values near neutrality. Enzymes function under conditions that support organic life.

REFERENCES

The information presented has been gleaned from a number of sources. The most useful of these are as follows:

James B. Sumner and G. Fred Somers, *Chemistry and Methods of Enzymes,* 2nd edition, Academic Press, New York, 1947.

Joseph S. Fruton and Sofia Simmonds, *General Biochemistry,* 2nd edition, Chapter 8, Wiley, New York, 1958.

Abraham White, Philip Handler, and Emil L. Smith, *Principles of Biochemistry,* 5th edition, Chapter 10, McGraw-Hill, New York, 1973.

CHEMICAL
KINETICS

Kinetics is the study of the velocities of chemical reactions. The purpose of kinetic studies is the deduction of the reaction mechanisms, that is, the identification of the intermediate steps by which reactants are converted into products. The study of kinetics proceeds in two major identifiable steps: (1) the collection of data and (2) the formulation of a theory in agreement with the data. In the following pages kinetic analysis is discussed from this point of view, starting with the simplest kinetic experiments. (If the reader believes that he has a firm grasp of the fundamentals, he should feel free to skip the next few pages.)

First-Order Reactions

The simplest chemical reaction is reactant A forming product P:

$$A \rightarrow P. \tag{1}$$

One may study the conversion of A to P by observing experimentally a change in some physical or chemical property which is characteristic of either the reactant A or the product P. Properties often measured include the absorbance of light by either reactant or product, the uptake of acid or base by the product, and the incorporation of a radioactive isotope originally associated with reactant into radioactive product. Many other properties have been used in kinetic studies, but these are presented as illustrative examples. In the treatments below, the concentration of reactant A is used as the experimental basis; however, any experimental quantity directly proportional to concentration may also be used. The data obtained experimentally may be used to show the change in concentration of the reactant A as a function of time (Fig. 2-1).

The rate of a reaction is defined as the decrease in the concentration of reactant A with time, $-d[A]/dt$, or the increase in the concentration of product P with time, $d[P]/dt$. For the sake of convenience, the former term generally will be used. A fundamental assumption of chemical kinetics is that the kinetic form of a one-step reaction is identical to its stoichiometric form. Therefore, the theory which explains the experimental data for the conversion of A to P states that the reaction rate is proportional to the concentration of A. By using a proportionality constant which is called the rate constant, k, one may say that the reaction rate is equal to k multiplied by the concentration of A, that is:

$$-\frac{d[A]}{dt} = k[A].$$
(2)

A theoretical expression such as this which describes the rate of a reaction as a function of reactant concentration is termed a *rate law*. At any given time the rate of the conversion of A to P, $-d[A]/dt$, is the slope of the tangent to the plot of the concentration of A as a function of time (Fig. 2-1). Obviously, the slope of the tangent to this plot varies with time. The rate constant, k, remains constant as the concentration of A and the time vary. The rate of the reaction is observed experimentally, while the rate constant is evaluated from the mathematical analysis of the kinetic study.

The rate constant which describes the conversion of A to P may be determined if Eq. (2) is converted to a simple algebraic ex-

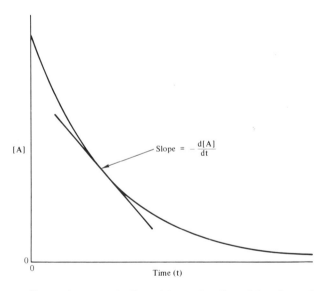

Fig. 2-1. Change in concentration of A as a function of time for a chemical reaction.

pression. This may be done by rearranging and integrating Eq. (2):

$$-\int \frac{d[A]}{[A]} = k \int dt, \tag{3}$$

or

$$- \ln [A] = kt + \text{constant}, \tag{4}$$

and

$$\log [A] = - (kt/2.303) - (\text{constant}/2.303). \tag{5}$$

Since Eq. (5) is a linear algebraic expression, a plot of the log of the concentration of A as a function of time yields a straight line (Fig. 2-2). The slope of this line is equal to $-k/2.303$; the intercept of the ordinate at time = 0 is $-$ constant/2.303.

In a kinetic study, a plot of the log of the concentration of A obtained experimentally versus time may yield a straight line. If this

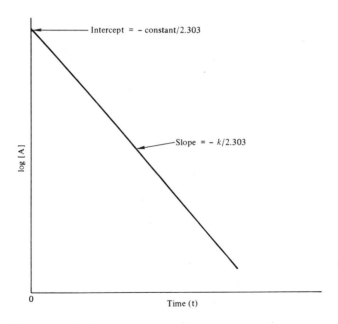

Fig. 2-2. Linear plot for a first-order reaction.

is the case, the conversion of A to P obeys the theoretical expression given by Eq. (2). Since the rate of the reaction is found experimentally to be proportional to the concentration of *one* reacting substance (in this case the concentration of A), the reaction is said to be first-order. Furthermore, the rate constant for this reaction may be evaluated by multiplying the slope of the plot of the log of the concentration of A versus time by -2.303. This first-order rate constant has the units of time^{-1}.

An alternative method may be used to convert Eq. (2) into an algebraic expression. Equation (2) may be rearranged and integrated between the limits of concentration $[A]_0$ at time $t = 0$ and $[A]$ at time $t = t$:

$$- \int_{[A]_0}^{[A]} \frac{d[A]}{[A]} = k \int_0^t dt, \tag{6}$$

$$- \ln \frac{[A]}{[A]_0} = kt, \tag{7}$$

and

$$\ln \frac{[A]_0}{[A]} = kt, \tag{8}$$

or

$$\log \frac{[A]_0}{[A]} = \frac{kt}{2.303}. \tag{9}$$

Once again, in a kinetic experiment a plot of $\log ([A]_0/[A])$ versus time may yield a linear plot (Fig. 2-3). If this is the case, the conversion of A to P obeys the rate expression given by Eq. (2); and therefore the reaction is first-order. Also, the rate constant, k, may be evaluated by multiplying the slope of this plot by 2.303.

Occasionally, it is found convenient to express the rate of a reaction in terms of its *half-life, $t_{1/2}$,* that is, the time required to reduce the concentration of A to half of its initial value. At $t_{1/2}$ for a first-order reaction, $[A] = [A]_0/2$, and Eq. (9) becomes

$$\log \frac{[A]_0}{[A]_0/2} = \frac{kt_{1/2}}{2.303}. \tag{10}$$

This expression can then be solved for $t_{1/2}$:

$$t_{1/2} = \frac{2.303 \log 2}{k} = \frac{0.693}{k}. \tag{11}$$

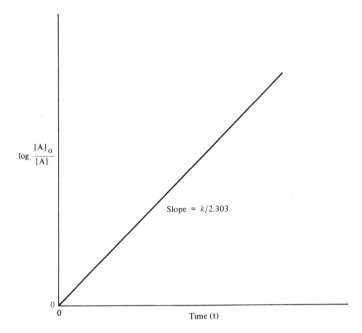

$$\log \frac{[A]_0}{[A]}$$

Slope $= k/2.303$

0

0

Time (t)

Fig. 2-3. Linear plot for a first-order reaction.

Thus, the half-life of a first-order reaction may be calculated from the first-order rate constant and vice versa by using Eq. (11).

Second-Order Reactions

One may conceive of a reaction in which two molecules of A react to yield product P:

$$2A \rightarrow P. \tag{12}$$

Again, it is assumed that the rate of such a reaction is proportional to its stoichiometric form, i.e., the square of the concentration of the reactant A:

$$- \frac{d[A]}{dt} = k[A][A] = k[A]^2. \tag{13}$$

15

This expression may be rearranged and integrated between the limits of concentration $[A]_0$ at time $t = 0$ and $[A]$ at time $t = t$:

$$-\int_{[A]_0}^{[A]} \frac{d[A]}{[A]^2} = k \int_0^t dt, \tag{14}$$

and

$$\frac{1}{[A]} - \frac{1}{[A]_0} = kt, \tag{15}$$

or

$$\frac{1}{[A]} = kt + \frac{1}{[A]_0}. \tag{16}$$

A plot of $1/[A]$ versus time is linear for this second-order reaction (Fig. 2-4). The slope is positive and equal to the second-order rate constant k; the intercept of the ordinate at time $= 0$ is $1/[A]_0$.

If, in a kinetic study, a plot of $1/[A]$ versus time is linear the conversion of A to P obeys the theoretical expression given by Eq. (13). Since the rate of the reaction is found experimentally to be proportional to the concentration of A to the second power, the reaction is second-order. The rate constant for a second-order reaction has the units of concentration^{-1} time^{-1}.

One may also conceive of a reaction in which two different compounds A and B react to form product P:

$$A + B \rightarrow P. \tag{17}$$

The second-order kinetic nature of such a reaction is suggested by its stoichiometry. A theoretical expression, or rate law, which describes the rate of this reaction is

$$-\frac{d[A]}{dt} = k[A][B]. \tag{18}$$

If A and B are present in equal concentrations, that is if $[A] = [B]$, then

$$-\frac{d[A]}{dt} = k[A][B] = k[A]^2. \tag{19}$$

The integrated rate expression given by Eq. (16) may be validly applied. Therefore, a plot of $1/[A]$ versus time or $1/[B]$ versus time is a straight line, the slope of which is equal to the second-order rate constant k, as in Fig. 2-4.

On occasions, half-lives are also used to describe the rates of

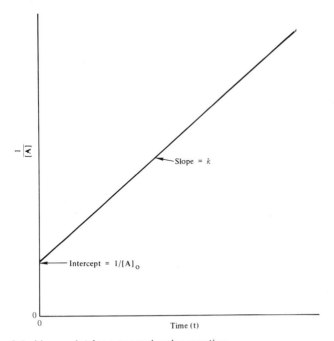

Fig. 2-4. Linear plot for a second-order reaction.

second-order reactions. By definition, at $t_{1/2}$, $[A] = [A]_0/2$. This may be substituted into Eq. (16) to give

$$\frac{1}{[A]_0/2} = kt_{1/2} + \frac{1}{[A]_0}. \tag{20}$$

This equation can then be solved for $t_{1/2}$:

$$t_{1/2} = \frac{1}{k[A]_0}. \tag{21}$$

Thus, by using Eq. (21), the half-life of a second-order reaction may be calculated from the second-order rate constant and $[A]_0$, and the rate constant can be calculated from the half-life and $[A]_0$.

An experiment in which two different compounds such as A and B react is often simplified by having one of the reactants, for example B, present in great excess over the other. The concentration of B will decrease only slightly during the reaction and is considered constant; and therefore the rate is proportional only to the

17

concentration of A. The observed reaction will obey the theoretical treatment for a first-order reaction and give an observed first-order rate constant, k_{obsd}. Since the reaction appears to follow first-order kinetics, but is in reality second-order, it is said to be a *pseudo* first-order reaction; the rate constant for this reaction, k_{obsd}, is a *pseudo* first-order rate constant. Qualitatively, the reaction will appear to follow the rate law,

$$-\frac{d[A]}{dt} = k_{obsd}[A].$$
(22)

Equations (18) and (22) may be combined to show that

$$k_{obsd}[A] = k[A][B],$$
(23)

and

$$k_{obsd} = k[B].$$
(24)

The second-order rate constant, k, may be evaluated by determining k_{obsd} at several different concentrations of B. A plot of k_{obsd} versus concentration of B is linear (Fig. 2-5, Line a), and the slope of this line is the second-order rate constant which has the units of concentration^{-1} time^{-1}.

Numerous reactions which occur in aqueous solution are catalyzed by weak acids or weak bases that are included in the solution as buffers. These reactions obey the rate law for a second-order reaction, Eq. (18), but are observed to be *pseudo* first-order. The second-order rate constant for catalysis by buffer may be determined as the slope of a plot of k_{obsd} versus concentration of buffer, B (Fig. 2-5, Line b). In such an experiment the intercept of the vertical axis corresponds to the first-order rate constant for that portion of the reaction which is not catalyzed by buffer. This portion of the total reaction may be catalyzed by the solvent species, hydrogen ion, hydroxide ion, and/or water itself. A detailed analysis of these reactions, which are catalyzed by solvent species and by weak acids and weak bases, is examined in detail in Chapter 4.

Third-Order Reactions

Third-order reactions are rare. Those that have been studied generally are of the type in which two molecules of A react with one molecule of B to yield product:

$$2A + B \rightarrow P. \tag{25}$$

The rate of this reaction is given by

$$-\frac{d[A]}{dt} = k[A]^2[B]. \tag{26}$$

Most of the reactions which follow this rate law have been observed in the gas phase. These reactions are peripheral to the subject of this text; consequently, they will not be treated further.

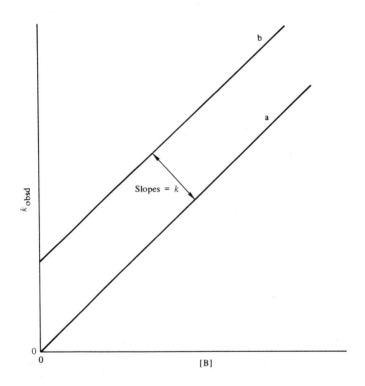

Fig. 2-5. Line a: Variation of the observed first-order rate constant, k_{obsd}, with concentration of B in a second-order reaction. Line b: Same as Line a except that at all concentrations of B there is a contribution to k_{obsd} due to catalysis by solvent species such as hydrogen ion, hydroxide ion, and/or water itself.

Kinetic Order and Molecularity

In the preceding pages the term *order* has been used to classify chemical reactions on the basis of their kinetic behavior. Kinetic order is defined as the sum of the exponents of the concentrations of reactants in the rate law. Each individual exponent is called the order with respect to that component. For example, in the third-order reaction described by the rate law

$$-\frac{d[A]}{dt} = k[A]^2[B],\tag{26}$$

the reaction is third-order since the sum of the exponents is three. This reaction is second-order with respect to reactant A since the exponent of its concentration is two; and it is first-order with respect to reactant B since the exponent of its concentration is one. The kinetic order of a reaction is derived experimentally, and it is an empirical concept.

The term *molecularity* describes the number of molecules of reactant which collide during the reaction process. The molecularity of a reaction is deduced from the empirical data, and it is a theoretical concept.

Dynamic Equilibrium

To this point, we have assumed that reactants are completely converted to products. However, this is the limiting case. Reactants are never completely converted to products because products have a tendency to re-form the reactants. Equilibria are set up between reactants and products. An equilibrium reaction may be generalized as

$$A + B \rightleftharpoons P + Q,\tag{27}$$

where A and B are reactant species, and P and Q are product species. The assumption which forms the basis for further mathematical treatment is that chemical equilibrium is a dynamic, and not a static, condition. At equilibrium, both the forward and the reverse reactions are taking place. However, the rates of these reactions balance, so that the concentrations of all reactants and products remain constant.

When viewed quantitatively, the rate of the forward reaction is given by

$$-\frac{d[A]}{dt} = k_{forward}[A][B], \tag{28}$$

and the rate of the reverse reaction is given by

$$-\frac{d[P]}{dt} = k_{reverse}[P][Q]. \tag{29}$$

At equilibrium, the rate of the forward reaction is equal to the rate of the reverse reaction, and

$$k_{forward}[A][B] = k_{reverse}[P][Q]. \tag{30}$$

This expression, when rearranged, yields

$$\frac{[P][Q]}{[A][B]} = \frac{k_{forward}}{k_{reverse}} = K. \tag{31}$$

By convention the concentrations of the products are placed in the numerator, and the concentrations of the reactants are placed in the denominator. The constant, K, is called the *equilibrium constant* for the reaction.

The existence of an equilibrium and the magnitude of the equilibrium constant, K, are functions of the differences in free energies of the reactants and products. If a reaction appears to go to completion, it reflects a large negative ΔG between reactants and products. If a reaction results in a detectable equilibrium with a measurable equilibrium constant, K, the difference in free energy between reactants and products is relatively small. Indeed, the magnitude of ΔG may be calculated from the equilibrium constant which describes the reaction (Chapter 3).

REFERENCES

Arthur A. Frost and Ralph G. Pearson, *Kinetics and Mechanism,* 2nd edition, Chapter 2, Wiley, New York, 1961.

William P. Jencks, *Catalysis in Chemistry and Enzymology,* Chapter 11, McGraw-Hill, New York, 1969.

PROBLEMS

1. The reactions described in this chapter assumed that reactant A is converted directly to product P without formation of an intermediate. Consider the bimolecular reaction in which reactants A and B react reversibly to form the intermediate X which then reacts to form P (enzymatic reactions proceed by a similar mechanism):

$$A + B \rightleftharpoons X \rightarrow P.$$

Graphically show how the concentrations of A,B,X, and P would be expected to vary as a function of time; assume equal initial concentrations of A and B.

2. The hydrolysis of phenylacetate at 5°C in 0.6 M ethylamine buffer, 40% of which was in the free base form, gave the accompanying data (adapted from W. P. Jencks, *Catalysis in Chemistry and Enzymology*, p. 559).

time (min)	[phenylacetate] (arbitrary units)
0	0.55
0.25	0.42
0.50	0.31
0.75	0.23
1.00	0.17
1.25	0.12
1.50	0.085

    a. What is the observed *pseudo* first-order rate constant for this reaction?

    b. What is the half-life of the reaction, $t_{1/2}$, under the conditions employed?

3. Phenacyl bromide and pyridine react in methyl alcohol to form the quaternary ammonium salt, phenacylpyridinium bromide:

$$C_6H_5-\overset{\overset{\textstyle O}{\|}}{C}-CH_2Br + C_5H_5N \rightarrow C_6H_5-\overset{\overset{\textstyle O}{\|}}{C}-CH_2\overset{\oplus}{N}C_5H_5 + Br^{\ominus}.$$

When equal concentrations of reactants were used at 35°C, the following data were obtained; these describe the time course of the reaction (adapted from A. A. Frost and R. G. Pearson, *Kinetics and Mechanism,* 2nd edition, p. 36).

time (min)	[reactant] (M)
0	.0385
100	.0330
200	.0288
300	.0255
400	.0230
500	.0208
600	.0191
700	.0176
800	.0163
1,000	.0143

a. What is the order of this reaction?
b. What is its rate constant?
c. What is the half-life of this reaction?

4. The hydrolysis of o-nitrophenyl bromoacetate is catalyzed by phosphate buffers (B. Holmquist and T. C. Bruice, *J. Am. Chem. Soc.* **91**, 2982 (1969)).

a. From the following data, calculate the second-order rate constant for this reaction.

[phosphate buffer] (M)	k_{obsd} (sec^{-1})
0.03	1.7
0.06	2.7
0.12	4.5
0.18	6.5
0.24	8.5
0.30	10.5

b. Why is it not valid to calculate the second-order rate constant by dividing the observed *pseudo* first-order rate constants by the buffer concentrations?

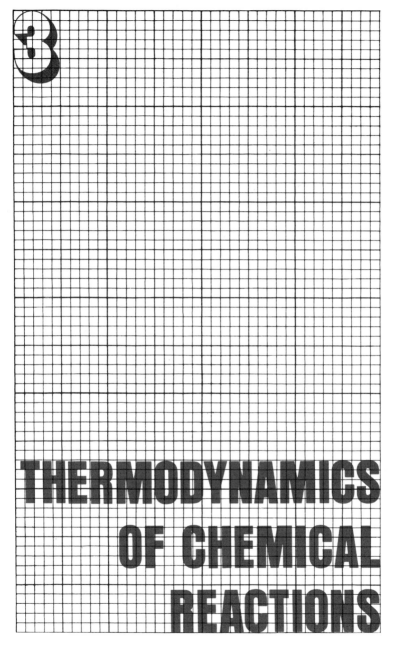

3

THERMODYNAMICS OF CHEMICAL REACTIONS

Chemical and enzyme-catalyzed reactions are governed by the flow of energy as reactants are converted to products. Both the possibility of a reaction taking place and the rate at which it takes place are functions of energetic considerations. These considerations have already been alluded to in the discussion of the nature of catalysis in Chapter 1. The science of thermodynamics contains the laws that describe energy changes in physical and chemical processes. The major concepts of this discipline are stated in two unifying principles, the first and second laws of thermodynamics. These laws allow one to describe the rate of a chemical reaction and its ultimate equilibrium position on the basis of energetic considerations. The following treatment of thermodynamics is not rigorous; however, it does cover the most important features.

The Laws of Thermodynamics

a. The First Law. Energetic transformations take place within regions of space. A *system* is the matter within a defined region. The *surroundings* encompass the remainder of the universe. A system is *open* if matter as well as energy in the form of heat or work can cross its boundaries. It is *closed* if matter cannot cross its boundaries but energy can. A system is *isolated* if neither matter nor energy can cross its boundaries.

The first law of thermodynamics states that energy can be neither created nor destroyed. In any process the total quantity of energy of the system and its surroundings remains constant. To illustrate: if, at the start of a process, the energy of a system is E_1, and at the end of the process the energy of the system has increased to E_2, the change in energy of the system, ΔE, is given by

$$\Delta E = E_2 - E_1. \tag{1}$$

The increase in energy of the system, ΔE, corresponds exactly to the loss of energy by the surroundings.

If a system receives heat from the surroundings, it may also do work. For example, a gas, when heated, will expand against atmospheric pressure. The change in energy of the system, ΔE, must then reflect the heat energy absorbed and the work energy expended. This change in energy of the system is given by

$$\Delta E = Q - W \quad \text{or} \quad Q = \Delta E + W \tag{2}$$

where Q is heat energy and W is work energy. By convention, when heat is absorbed, the numerical value given to Q is positive; when heat is lost, the value of Q is negative. When work is done on the system, the value of W is negative; when the system does work upon its surroundings, the value of W is positive. An important point to note about Eqs. (1) and (2) is that the change in energy of the system, ΔE, depends only on the initial and final states; the pathway taken during the transformation has no effect on ΔE.

Most chemical and enzymatic reactions proceed at constant (atmospheric) pressure. For a system operating under this condition, an absorption of heat from the surroundings $(+ Q)$ is accompanied by an increase of its volume; the system thereby performs work $(+ W)$. The work done at constant pressure, P, in increasing the volume is given by

$$W = P\Delta V \tag{3}$$

where ΔV is the difference in volume between the initial and final states. Equation (3) when substituted into Eq. (2) yields

$$Q = \Delta E + P\Delta V. \tag{4}$$

For a system operating at a constant pressure the change in heat energy, Q, is denoted by ΔH. The term H is defined as the *enthalpy* or the total *heat content* of the system; ΔH is the change in enthalpy or the change in heat content between the initial and final states. Thus, for a system operating at constant pressure, Eq. (4) can be rewritten as

$$\Delta H = \Delta E + P\Delta V, \tag{5}$$

where ΔH is the change in enthalpy of the system, ΔE is the change in total energy of the system, and $P\Delta V$ is the work done by the system upon its surroundings. While the terms ΔE and $P\Delta V$ are significant in describing many chemical reactions, they are not commonly used in reference to reactions occurring in aqueous solutions. Such reactions do not undergo significant changes in volume. Hence, $P\Delta V$ is approximately equal to zero; and ΔH is approximately equal to ΔE. Thus, for a reaction occurring in aqueous solution the change in enthalpy, ΔH, adequately describes the change in total energy of the system. This term can be determined, and it is commonly used.

b. The Second Law of thermodynamics states that all systems

spontaneously tend to approach a state of equilibrium. A displacement of a system away from a state of equilibrium takes place only at the expense of energy derived by displacing another system toward equilibrium. This may be illustrated with a few simple examples. Water will flow downhill until it reaches its lowest possible level, the ocean. Water can be displaced back to the mountains by evaporation only at the expense of solar radiation. A watch will run down; it can be wound only by the input of mechanical energy. A match will burn, thereby releasing heat energy, but the reverse process is never seen.

In broad terms, the second law indicates a direction for the universe. It states that in all processes some energy becomes unavailable to participate in further processes. A portion of the enthalpy, or heat content of the system, ΔH, can no longer perform useful work since, in most cases, it has increased the random motions of molecules in the system. By definition

$$Q = T\Delta S \tag{6}$$

where Q is the total energy which loses its ability to do work, and T is the absolute temperature which is a measure of the speed of random molecular motions. S is termed the *entropy,* and it is a measure of randomness or disorder of the system at constant temperature. By rearranging Eq. (6), one can express the change in entropy of a system for any process as

$$\Delta S = \frac{Q}{T}. \tag{7}$$

ΔS is the difference in entropy levels of the initial and final states of the system.

In mathematical terms, the second law states that for a process to occur spontaneously, the sum of the entropies of the system and the surroundings must increase. That is,

$$\Delta S_{system} + \Delta S_{surroundings} > 0. \tag{8}$$

It should be noted that a spontaneous reaction within a given system may be accompanied by a decrease in entropy, but this decrease in entropy of the system is more than counterbalanced by an increase in entropy of the surroundings. If there is no exchange of energy between the system and its surroundings, i.e., the system is isolated, a spontaneous reaction within the system is always accompanied by an increase in entropy.

On a practical level, entropy is not used as a criterion for determining whether a process can occur spontaneously; it is not readily measured. To circumvent this difficulty, Gibbs devised the concept of free energy, G (also F in the older literature), which is of greater utility in determining whether a process can occur spontaneously. The fundamental principle states that the enthalpy, H, is the sum of the energy which is free to do work, G, and that which is not free to do work, $T \cdot S$, i.e.,

$$H = G + T \cdot S. \tag{9}$$

For any transformation in the system, ΔH, ΔG, and ΔS are the changes in enthalpy, free energy, and entropy, respectively, between the initial and final states. Thus, for any process the free energy relationship, Eq. (9), can be expressed as

$$\Delta H = \Delta G + T\Delta S. \tag{10}$$

This equation is generally expressed in the form in which ΔG has been solved for:

$$\Delta G = \Delta H - T\Delta S. \tag{11}$$

For a process occurring within an isolated system, there is no net change in heat content of the system. That is, $\Delta H = 0$, and

$$\Delta G = -T\Delta S. \tag{12}$$

It has been stated previously that for a reaction to be able to occur spontaneously, the net entropy change of the system and the surroundings must be positive, Eq. (8). This statement also holds true for an isolated system. Therefore, according to Eq. (12), for a system and its surroundings or an isolated system at constant temperature, a spontaneous reaction is characterized by a positive value for ΔS and a negative value for ΔG.

Evaluation of Differences in Thermodynamic Parameters Between Reactants and Products

To understand how an enzyme may effect catalysis of a chemical reaction, it is useful to have knowledge of the energy levels of reactants, transition state, and products. This point was illustrated in Chapter 1 with the reaction coordinate diagram. While it is difficult to measure absolute values of the enthalpy, free energy, and entropy, it is possible to measure differences in these quantities between points along the reaction pathway. There are experimental means which can be used to measure the differences in thermodynamic parameters, ΔH, ΔG, and ΔS, between reactants and products. There are also methods of evaluating the thermodynamic activation parameters $\Delta H^{\ddagger}$, $\Delta G^{\ddagger}$, and $\Delta S^{\ddagger}$.

a. ΔH. The difference in enthalpies between reactants and products of an irreversible reaction may be measured calorimetrically. For example, glucose may react with oxygen to yield carbon dioxide and water:

$$C_6H_{12}O_6 + 6O_2 \rightarrow 6H_2O + 6CO_2. \tag{13}$$

Values of enthalpies are expressed in units of calories per mole, where a calorie is the quantity of heat required to raise the temperature of one gram of water from 14.5 to 15.5°C. For the oxidation of glucose, Eq. (13), the change in enthalpy, ΔH, is $-673,000$ cal/mole.

Since the quantitative value of the change in enthalpy for a reaction, ΔH, and the other thermodynamic parameters vary slightly with conditions, it is useful to determine these values under *standard conditions*. The changes in the various parameters under standard conditions are expressed as $\Delta H°$, $\Delta G°$, and $\Delta S°$. They represent changes that would be observed when the reactants were in their standard states. For substances in solution, the standard state means a concentration (or more rigorously, an activity) of one mole/liter at 25°C.

The standard enthalpy change, ΔH^0, of a reversible chemical reaction may be calculated from the equilibrium constants for that reaction at different temperatures. The relationship between the change in equilibrium constant, K, as a function of temperature and the standard enthalpy of reaction, ΔH^0, is given by the van't Hoff equation:

$$\frac{d \ln K}{dT} = \frac{\Delta H^0}{RT^2} \tag{14}$$

where R is the gas constant (1.987 cal mole^{-1} deg^{-1}) and T is the absolute temperature. When integrated, this equation yields

$$\ln K = C - \frac{\Delta H^0}{RT}, \tag{15}$$

or

$$\log K = \frac{C}{2.303} - \frac{\Delta H^0}{2.303\,RT} \quad . \tag{16}$$

When log K is plotted versus the reciprocal of the absolute temperature, $1/T$, a linear plot is obtained (Fig. 3-1). The intercept of the vertical axis is the integration constant, C, divided by 2.303; the slope of the line is $-\Delta H^0/2.303\,R$. Enthalpy changes accompanying reversible reactions such as ionization equilibria can be evaluated by this method; and, as one would expect, they are much smaller than ΔH^0 values for irreversible reactions such as the oxidation of glucose, Eq. (13). Typical ΔH^0 values for the ionization of some groups found in proteins are: carboxyl, $0 \pm 1{,}500$ cal/mole; and amino, $+10{,}000$ to $+13{,}000$ cal/mole.

b. ΔG. The differences of free energy levels between reactants and products of a reversible reaction may also be calculated from the equilibrium constant. To begin, the free energy of any substance in solution at a given state, G, is related to its free energy in the standard state by

$$G = G^0 + RT \ln [A] \tag{17}$$

where R is the gas constant, T is the absolute temperature, and $[A]$ is the concentration (or, more rigorously, the activity) of the substance. For the reversible chemical reaction

$$A + B \overset{K}{\rightleftharpoons} C + D, \tag{18}$$

the change in free energy is given by

$$\Delta G = \Delta G^0 + RT \ln \frac{[C][D]}{[A][B]} \tag{19}$$

where $\Delta G^0 = G^0_C + G^0_D - G^0_A - G^0_B$. At equilibrium the reaction has gone to completion; the reactants have lost their ability to perform useful work, i.e., $\Delta G = 0$. Thus, Eq. (19) simplifies to

$$\Delta G^0 = -RT \ln \frac{[C][D]}{[A][B]} = -RT \ln K. \tag{20}$$

The standard free energy change, ΔG^0, of a reversible chemical reaction can be calculated from the equilibrium constant; like enthalpy, it has units of calories per mole. If ΔG^0 is negative and reactants are at their standard states, the process may proceed

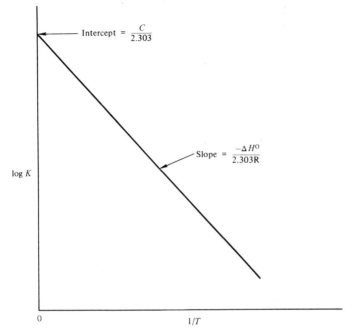

Fig. 3-1. A van't Hoff plot of the logarithm of the equilibrium constant of a reversible reaction versus the reciprocal of the absolute temperature at which it was determined.

spontaneously in the direction indicated. If ΔG^0 is positive and reactants are at their standard states, the net reaction is not thermodynamically feasible. It may take place only if free energy is made available to the system from the surroundings.

c. ΔS. The free energy relationship, as we have seen, contains three terms:

$$\Delta G = \Delta H - T\Delta S. \tag{11}$$

Since, for a chemical reaction, the changes in enthalpy, ΔH, and free energy, ΔG, can be determined experimentally, the change in entropy at a given absolute temperature, ΔS, can be calculated simply from Eq. (11). Entropy has units of calories per mole per degree; these units are often abbreviated e.u. for entropy units.

33

Evaluation of Thermodynamic Activation Parameters

a. E_a. The rates of chemical and enzyme-catalyzed reactions are a function of temperature. The rate of a reaction generally increases from two- to fourfold with a temperature increase of 10°C. In 1889 Arrhenius showed that the rate constant of a reaction increases in an exponential manner with the temperature. The empirical Arrhenius relationship is

$$\frac{d \ln k}{dt} = \frac{E_a}{RT^2}, \tag{21}$$

where k is the rate constant of the reaction under study, R is the gas constant, T is the absolute temperature, and E_a is energy of activation. Upon integration, the Arrhenius equation (21) yields

$$\ln k = -\frac{E_a}{RT} + \ln A, \tag{22}$$

or

$$\log k = -\frac{E_a}{2.303\,RT} + \log A. \tag{23}$$

If $\log k$ for a given reaction is plotted versus the reciprocal of the absolute temperature, $1/T$, a linear plot may be obtained (Fig. 3-2). The intercept of the vertical axis is the integration constant, $\log A$; the slope of the line is $-E_a/2.303\,R$. E_a has units of calories per mole.

To explain the temperature dependence of reaction rate constants, Arrhenius postulated that reactants must first be converted to a high-energy species, which we now call the *activated complex,* and that this activated complex decomposes to products. The energy required by the system to convert reactants to the activated complex is the Arrhenius energy of activation, E_a. While this theory is useful in explaining the temperature dependence of reactions, it does not explain reaction rates in the common thermodynamic terms of enthalpy, H, entropy, S, or free energy, G. That explanation comes from transition-state theory.

b. $\Delta H^{\ddagger}$. The *transition-state theory* was developed largely through the efforts of Eyring to more accurately describe the velocity of a chemical reaction in thermodynamic terms. It holds that the reactant or reactants must first reach a transition state (i.e., activated complex), and that the rate of reaction is propor-

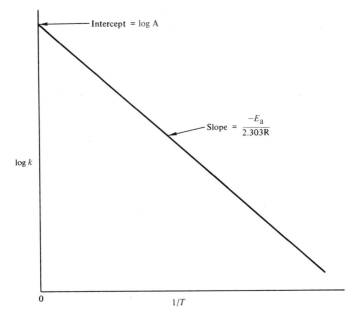

Fig. 3-2. An Arrhenius plot of the log of rate constant versus the reciprocal of the absolute temperature at which it was determined.

tional to the concentration of this activated complex. Let us consider, for example, a simple bimolecular reaction:

$$A + B \underset{K^{\ddagger}}{\overset{K^{\ddagger}}{\rightleftharpoons}} A \cdot B^{\ddagger} \overset{k^{\ddagger}}{\rightarrow} \text{Product.} \tag{24}$$

According to the transition-state theory, the rate of reaction is given by

$$\frac{-d[A]}{dt} = k^{\ddagger} [A \cdot B^{\ddagger}]. \tag{25}$$

Since the equilibrium constant for activated complex formation is given by

$$K^{\ddagger} = \frac{[A \cdot B^{\ddagger}]}{[A][B]}, \tag{26}$$

we may solve for $[A \cdot B^{\ddagger}]$:

$$[A \cdot B^{\ddagger}] = K^{\ddagger}[A][B]. \tag{27}$$

35

By substituting Eq. (27) for $[A \cdot B^{\ddagger}]$ in Eq. (25), we can demonstrate that

$$\frac{-d[A]}{dt} = k^{\ddagger} K^{\ddagger} [A][B].$$

(28)

This equation has the form of the rate law for a second-order reaction in which the observed rate constant is given by

$$k = k^{\ddagger} K^{\ddagger}.$$

(29)

The rate constant of breakdown of the activated complex, $k^{\ddagger}$, can be estimated on theoretical grounds. One first assumes that the rate constant of breakdown, $k^{\ddagger}$, is equal to the vibrational frequency, ν, that leads to decomposition. (To be precise, ν should be multiplied by the *transmission coefficient, κ*, which is the probability that the activated complex will decompose to products; this is usually assumed to be unity, and therefore omitted.) According to physical principles, the frequency, ν, is equal to ϵ/h where ϵ is the average energy of vibration, and h is Planck's constant. Therefore,

$$k^{\ddagger} = \nu = \frac{\epsilon}{h}.$$

(30)

At the temperature, T, an excited vibration has the classical energy $\epsilon = k_B T$ where k_B is Boltzmann's constant. By substituting this value of ϵ into Eq. (30), it can be shown that

$$k^{\ddagger} = \frac{k_B T}{h},$$

(31)

and the rate constant is given by

$$k = \frac{k_B T}{h} K^{\ddagger}.$$

(32)

The equilibrium for the formation of the activated complex can be described by the common thermodynamic terms. Therefore, it follows that

$$\Delta G^{\ddagger} = -RT \ln K^{\ddagger}.$$

(33)

$K^{\ddagger}$ may then be solved for:

$$K^{\ddagger} = e^{(-\Delta G^{\ddagger}/RT)}.$$

(34)

The rate constant for the reaction may then be described by substituting this value for $K^{\ddagger}$ into Eq. (32) to yield

$$k = \frac{k_B T}{h} e^{\,(-\Delta G^{\ddagger}/RT)}.$$ (35)

Again, since formation of the activated complex can be described by the standard thermodynamic terms,

$$\Delta G^{\ddagger} = \Delta H^{\ddagger} - T\Delta S^{\ddagger}.$$ (36)

The reaction rate constant, k, may then be given by

$$k = \frac{k_B T}{h} e^{\,(-\Delta G^{\ddagger}/RT)} = \frac{k_B T}{h} e^{\,(-\Delta H^{\ddagger}/RT)} e^{\,(\Delta S^{\ddagger}/R)}.$$ (37)

If one assumes that $\Delta S^{\ddagger}$ does not vary with temperature, one may take the logarithm of Eq. (37) and differentiate it to obtain

$$\frac{d \ln k}{dt} = \frac{1}{T} + \frac{\Delta H^{\ddagger}}{RT^2},$$ (38)

or

$$\frac{d \ln k}{dt} = \frac{\Delta H^{\ddagger} + RT}{RT^2}.$$ (39)

This equation, which is derived on the basis of the assumptions of transition-state theory and which describes rate constants in terms of $\Delta H^{\ddagger}$, is similar in form to the empirical Arrhenius relationship, Eq. (21). By equating these two, one can show that

$$\frac{E_a}{RT^2} = \frac{\Delta H^{\ddagger} + RT}{RT^2},$$ (40)

and

$$E_a = \Delta H^{\ddagger} + RT,$$ (41)

or

$$\Delta H^{\ddagger} = E_a - RT.$$ (42)

Thus, transition-state theory bridges the empirical observations of Arrhenius with thermodynamics. Among other things, it gives a method of correcting the activation energy, E_a, which is obtained experimentally to give an actual enthalpy of activation by the use of Eq. (42). This correction of activation energy to obtain the enthalpy of activation is often significant, since under standard conditions (molar concentrations and 25°C) RT is equal to 592.4 cal/mole.

c. $\Delta G^{\ddagger}$. The rate of a chemical reaction is a function of the equilibrium formation of a transition-state species, or activated complex, Eq. (25). Therefore the free energy of activation, $\Delta G^{\ddagger}$, may be calculated from rate data in a manner analogous to the calculation of ΔG for a reversible reaction from equilibrium data, Eq. (20):

$$\Delta G^{\ddagger} = -RT \ln K^{\ddagger}, \tag{33}$$

where $K^{\ddagger}$ is the equilibrium constant for formation of the transition state. $K^{\ddagger}$ can be solved for from Eq. (32) as

$$K^{\ddagger} = \frac{kh}{k_B T}. \tag{43}$$

This value may then be substituted into Eq. (33) to give

$$\Delta G^{\ddagger} = -RT \ln \frac{kh}{k_B T}, \tag{44}$$

where k is the rate constant at temperature T, h is Planck's constant (1.584×10^{-34} cal·sec), and k_B is Boltzmann's constant (3.298×10^{-24} cal/deg). To keep units in agreement when calculating $\Delta G^{\ddagger}$ from Eq. (44), rate constants must be expressed in units of inverse time, sec^{-1}; and at 25°C the free energy of activation can be calculated as

$$\Delta G^{\ddagger} = -1,360 \log k + 17,400 \text{ cal/mole}. \tag{45}$$

d. $\Delta S^{\ddagger}$. The free energy relationship, Eq. (11), is valid when applied to the thermodynamic parameters of activation:

$$\Delta G^{\ddagger} = \Delta H^{\ddagger} - T \Delta S^{\ddagger} \tag{36}$$

Since $\Delta G^{\ddagger}$ and $\Delta H^{\ddagger}$ for a given reaction may be obtained from kinetic data, the calculation of the entropy of activation at a constant temperature, $\Delta S^{\ddagger}$, may be solved for in a simple mathematical exercise by using Eq. (36).

The Physical Meaning of the Thermodynamic Activation Parameters

As we have seen, thermodynamic parameters, especially the free energy of reaction, ΔG, and the free energy of activation, $\Delta G^{\ddagger}$, may be used to construct a reaction coordinate diagram (Chapter 1).

This diagram has utility in illustrating, in energetic terms, how reactions take place. However, this type of treatment is more quantitative than conceptual. It describes energetic changes without considering the physical meaning of these changes on a molecular level. It is important to understand how the thermodynamic activation parameters are related to reaction mechanisms.

The free energy of activation, $\Delta G^{\ddagger}$, is directly related to the reaction rate, Eq. (44), and it is the sum of those energetic factors which influence the reaction rate, Eq. (36). The enthalpy of activation, $\Delta H^{\ddagger}$, is a measure of the energy barrier which must be overcome by reacting molecules. It quantitatively describes the amount of heat energy the reacting molecules must acquire while being transformed from the energy level of reactant to that of the transition state.

The entropy of activation, $\Delta S^{\ddagger}$, is a measure of the fraction of reactants with sufficient activation enthalpy to react which can actually react; it includes factors such as concentration and solvent effects, steric requirements, and orientation requirements. If these factors come into play, they are reflected in large negative values of $\Delta S^{\ddagger}$, which in turn raise the value of $\Delta G^{\ddagger}$, which in turn results in a decrease of the observed reaction rate.

To illustrate this point, unimolecular reactions generally have entropies of activation near zero. If the molecules acquire sufficient energy to react (i.e., they absorb an amount of energy equal to $\Delta H^{\ddagger}$), they will react; they will not need to orient themselves in three-dimensional space. Multimolecular reactions have negative entropies of activation. Among other factors, negative entropies are the result of the requirement of reacting molecules to orient themselves properly in three-dimensional space, and to achieve an appropriate spatial proximity before reaction can take place. In simple terms, a negative entropy change implies an increase in order of a system, and a negative activation entropy implies that the reacting molecules must order themselves in the transition state before they may proceed to products. In general, then, unimolecular reactions have $\Delta S^{\ddagger}$ values which are near zero or positive; bimolecular reactions have negative $\Delta S^{\ddagger}$ values ranging from -5 e.u. to approximately -40 e.u.

By examining the thermodynamic activation parameters of enzyme-catalyzed reactions we can begin to understand how enzymes effect catalysis. For the purpose of illustration we will consider a single enzyme, rhodanase, which catalyzes the reaction of cyanide ion with thiosulfate to yield thiocyanate and sulfite:

$$CN^- + S—SO_3^{-2} \rightarrow SCN^- + SO_3^{-2} \tag{46}$$

The thermodynamic activation parameters of this reaction, when it occurs spontaneously and when it is catalyzed by rhodanase, are summarized in Table 3-1. The reaction, when catalyzed by the enzyme, has a much faster rate and consequently a free energy of activation, $\Delta G^{\ddagger}$, 8.6 kcal/mole less than the uncatalyzed reaction. According to Eq. (36) the decrease in $\Delta G^{\ddagger}$, for the enzymatic reactions could be the result of a decrease in the activation enthalpy, $\Delta H^{\ddagger}$, or an increase in the activation entropy, $\Delta S^{\ddagger}$, or both. An examination of the activation parameters of the uncatalyzed and enzyme-catalyzed reactions (Table 3-1) indicates that $\Delta H^{\ddagger}$ is lowered and $\Delta S^{\ddagger}$ is raised. Rhodanase lowers the activation enthalpy for the reaction, $\Delta H^{\ddagger}$, by 4.8 kcal/mole; therefore, the reactants require that much less heat energy to reach the transition state as they react. The enzyme raises the entropy of activation, $\Delta S^{\ddagger}$, a very substantial 12.8 cal/deg/mole.

Interpretation of the increased entropy of activation, $\Delta S^{\ddagger}$, on a molecular level requires knowledge of the rate-limiting step of the enzymatic reaction. This step, which dictates the rate of the overall reaction and from which the activation parameters are calculated, may be either the actual chemical reaction or conformational changes in the protein which precede or follow the bond-forming and bond-breaking steps. The former is the case if the substrate is poor or other experimental conditions dictate that bond formation or cleavage is rate-limiting. In such circumstances the

Table 3-1. Comparison of activation parameters for reactions of thiosulfate with cyanide[a]

Parameter[b]	Uncatalyzed	Catalyzed by rhodanase	Difference
$\Delta G^{\ddagger}$	23.7 kcal/mole	15.1 kcal/mole	8.6 kcal/mole
$\Delta H^{\ddagger}$	12.3 kcal/mole	7.5 kcal/mole	4.8 kcal/mole
$\Delta S^{\ddagger}$	−38.0 e.u.	−25.2 e.u	−12.8 e.u

[a] Data from Karen Leininger and John Westley, *J. Biol. Chem.* **243**, *1892–1899* (1968).
[b] Parameters measured at 300°K.

lowered entropy of activation indicates that the reactants require less ordering in going to the transition state than they would in the uncatalyzed reaction. Thus, the mechanism of the enzymatic reaction would impose upon the reactants its own orientation and steric requirements, a different means of achieving spatial proximity of the reactants, and its own solvent and concentration effects. However, it is now becoming clear that the rate-limiting steps of many enzymatic reactions are conformational changes which precede or follow the actual catalytic step. In these cases the activation entropy, $\Delta S^{\ddagger}$, reflects changes of the ordering of the structure of the enzyme itself during catalysis.

REFERENCES

Gordon M. Barrow, *Physical Chemistry,* 2nd edition, Chapter 15, McGraw-Hill, New York, 1966.

Malcolm Dixon and Edwin C. Webb, *Enzymes,* 2nd edition, Chapter 4, Academic Press, New York, 1964.

Arthur A. Frost and Ralph G. Pearson, *Kinetics and Mechanism,* 2nd edition, Chapter 5, Wiley, New York, 1961.

William P. Jencks, *Catalysis in Chemistry and Enzymology,* Chapter 11, McGraw-Hill, New York, 1969.

Edward S. West, Wilbert R. Todd, Howard S. Mason, and John T. Van Bruggen, *Textbook of Biochemistry,* 4th edition, Chapter 20, Macmillan, New York, 1966.

Abraham White, Philip Handler, and Emil L. Smith, *Principles of Biochemistry,* 5th edition, Chapters 10 and 11, McGraw-Hill, New York, 1973.

PROBLEMS

1. Imidazole and its monoprotonated form exist in equilibrium:

The pK_a values of this dissociation measured at several temperatures are:

T (°C)	pK_a
10	7.54
18	7.36
25	7.22
34	7.02

Calculate ΔH^0, ΔG^0, and ΔS^0 for this ionization.

2. The hydrolysis of compound I

occurs spontaneously in water (D. Piszkiewicz and T. C. Bruice, *J. Am. Chem. Soc.* **90**, 2156 (1968)). The rate constants for this reaction at several temperatures are summarized below.

T (°C)	$k_{obsd}(\text{sec}^{-1})$
25	7.95×10^{-8}
30	2.37×10^{-7}
56.2	1.04×10^{-5}
78.2	1.45×10^{-4}

a. Calculate $\Delta H^{\ddagger}$, $\Delta G^{\ddagger}$, and $\Delta S^{\ddagger}$ for this reaction at 25°C.

b. Is the rate-limiting step of this reaction likely to be unimolecular (i.e., involve only compound I) or bimolecular (i.e., involve compound I and water)?

3. *p*-Methylphenyl acetate reacts with imidazole to produce *p*-methylphenol and acetyl imidazole. The rate constants for this second-order reaction at a series of temperatures are as follows (data from T. C. Bruice and S. J. Benkovic, *J. Am. Chem. Soc.* **86**, 418 (1964)):

T (°C)	k (mole^{-1} sec^{-1})
10	2.34×10^{-2}
18	3.25×10^{-2}
25	4.5×10^{-2}
35	5.83×10^{-2}
42	7.5×10^{-2}
60	1.52×10^{-1}

Calculate $\Delta H^{\ddagger}$, $\Delta G^{\ddagger}$, and $\Delta S^{\ddagger}$, for this reaction.

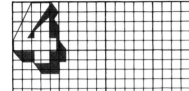

CATALYSIS IN AQUEOUS SOLUTIONS

Life on this earth originated in the primordial sea. The chemical reactions of the evolutionary process occurred in a watery environment; those that sustain living organisms continue to take place in an aqueous milieu. Water is essential to life, and its chemical properties are central to many enzymatic reactions and chemical reactions of biological significance. While it is not the purpose of this text to review the chemistry of water, it is appropriate to review briefly a few of the properties of water which are relevant to the kinetics of chemically and enzyme-catalyzed reactions. (This fundamental information is present for completeness. If the reader has a firm grasp of these concepts, he should feel free to skip the next few pages.)

Water, Acids, and Bases

Water does not exist solely as the molecular species H_2O, but dissociates slightly to hydrogen ion and hydroxide ion:

$$H_2O \rightleftharpoons H^+ + OH^-. \tag{1}$$

The equilibrium constant for this dissociation is

$$K_a = \frac{[H^+][OH^-]}{[H_2O]}. \tag{2}$$

The concentration of water (55.5 M) is not changed significantly by this ionization. Therefore, Eq. (2) is commonly simplified as

$$K_w = [H^+][OH^-], \tag{3}$$

where $K_w = 55.5\ MK_a$; the term K_w is the ion product of water and has a value of 10^{-14} M at 25°C. It is worth noting that according to Eq. (3) the concentrations of hydrogen ion and hydroxide ion are related in a reciprocal manner. When $[H^+]$ is high, $[OH^-]$ is low, and vice versa.

The definition of hydrogen ion and hydroxide ion concentrations in aqueous solutions is fundamental to chemistry. This is done with the conventional notation of pH which is a function of hydrogen ion concentration. By taking the logarithm of Eq. (3) and multiplying by -1, one can show that

$$-\log K_w = -\log [H^+] - \log [OH^-] = 14. \tag{4}$$

The function "$-\log$" is designated by "p"; hence, $-\log K_w = pK_w$,

$-\log[H^+] = pH$, and $-\log[OH^-] = pOH$. The notation pK_w is used occasionally; pOH is almost never used. pH is used universally; its definition ($pH = -\log[H^+]$), and significance should be understood implicitly.

Substances other than water dissociate to yield hydrogen ions:

$$BH \rightleftharpoons H^+ + B^-. \tag{5}$$

These substances such as BH which can give up protons or donate protons are *Brønsted acids.* Other substances such as the anion B^- accept a proton; these are *Brønsted bases.* Ionization of an acid such as BH yields its *conjugate base,* the anion B^-. Conversely, protonation of a base such as B^- yields its conjugate acid, BH.

The equilibrium constant for the dissociation of the acid BH is

$$K_a = \frac{[B^-][H^+]}{[BH]}. \tag{6}$$

This may be rearranged to

$$[H^+] = \frac{K_a[BH]}{[B^-]}. \tag{7}$$

By taking the logarithm of Eq. (7) and multiplying by -1, one obtains

$$-\log [H^+] = -\log K_a + \log [B^-] - \log [BH]. \tag{8}$$

Since $-\log [H^+] = pH$, and if $-\log K_a$ is defined as pK_a, Eq. (8) becomes the Henderson-Hasselbalch equation:

$$pH = pK_a + \log \frac{[B^-]}{[BH]}. \tag{9}$$

This equation defines the relationship between pH and the ratio of conjugate base to acid. It also defines the important term $pK_a = -\log K_a$, which is a measure of the strength of a substance as a Brønsted acid. The lower the value of pK_a, the stronger the substance is as an acid, and vice versa.

A common chemical experiment is the neutralization or titration of an acid with hydroxide ion.

$$BH + OH^- \rightarrow B^- + H_2O. \tag{10}$$

The data from such an experiment are commonly plotted as the equivalents of hydroxide added, which equals the equivalents of anion released, versus the pH (Fig. 4-1). The titration curve obtained has a characteristic sigmoid or S shape. For any point along this curve the Henderson-Hasselbalch equation (9) may be used to calculate the pH, the pK_a, or the ratio of conjugate base to acid if any two of these three factors are known. Also, it should be noted that if the acid is half-titrated, that is, if acid concentration, [BH], equals anion concentration, [B⁻], the Henderson-Hasselbalch equation simplifies to pH = pK_a. Thus, at the inflection point of a titration curve, the pH is equal to the pK_a of the acid.

Catalysis by Hydrogen Ion, Hydroxide Ion, and Water

The catalytic abilities of enzymes vary markedly with pH. In general enzymes exhibit a maximum catalytic activity at an intermediate pH; however, this optimum pH varies from enzyme to

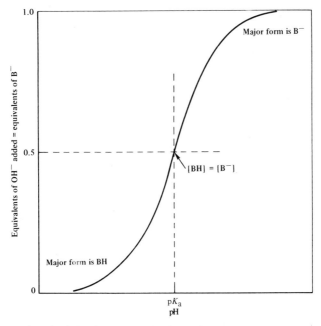

Fig. 4-1. A typical titration curve of a Brønsted acid, BH.

47

enzyme. By examining the dependence of the velocity on pH we gain an insight into the nature of the bond-forming and bond-breaking steps of the enzyme-catalyzed reaction. Before we can hope to understand how the pH dependence of an extremely complex catalyst such as an enzyme may be related to its mechanism, we must first understand how the pH dependence of simple chemical reactions may be related to their mechanisms.

Chemical reactions in aqueous solutions are often found to be dependent upon the concentration of hydrogen ion, hydroxide ion, and water itself. If these species are regenerated, they are by definition catalysts of the reaction. Reactions in which the catalytic species are ions related to the solvent are defined as being subject to *specific* catalysis. Thus, hydrogen ion is a *specific acid* catalyst in water, and hydroxide ion is a *specific base* catalyst in water. Many reactions are also dependent upon the concentrations of weak acids and bases in aqueous solution. These proton donors and proton acceptors, Brønsted acids and bases, are defined as *general acid* and *general base* catalysts. General catalysts are discussed in separate sections below.

Catalysis of a chemical reaction by hydrogen ion is a common phenomenon. A specific acid-catalyzed reaction can be generalized as

$$A + H^+ \rightarrow Product + H^+. \tag{11}$$

The rate of the reaction is given by

$$-\frac{d[A]}{dt} = k_H[H^+][A]. \tag{12}$$

Since the specific acid catalyst, hydrogen ion, is not consumed, its concentration remains constant throughout the reaction. The reaction appears to follow first-order kinetics, and a *pseudo* first-order rate constant, k_{obsd}, can be calculated. At any constant pH, the reaction will appear to follow the rate law,

$$-\frac{d[A]}{dt} = k_{obsd}[A]. \tag{13}$$

Equations (12) and (13) may be combined to show that

$$k_{obsd}[A] = k_H[H^+][A], \tag{14}$$

and

$$k_{obsd} = k_H[H^+].$$ (15)

The observed first-order rate constant, k_{obsd}, of a specific acid-catalyzed reaction varies linearly with hydrogen ion concentration. However, the commonly used measure of hydrogen ion concentration, pH, is $-\log[H^+]$. Consequently, a plot of k_{obsd} versus pH gives a curved line (Fig. 4-2**a**) in which k_{obsd} increases rapidly as pH decreases.

A common and far more useful method of presenting experimental results is to plot $\log k_{obsd}$ as a function of pH (Fig. 4-2**b**). The log of Eq. (15) is,

$$\log k_{obsd} = \log k_H + \log[H^+],$$ (16)

and

$$\log k_{obsd} = \log k_H - pH.$$ (17)

Clearly, a plot of $\log k_{obsd}$ versus pH gives a straight line (Fig. 4-2**b**). The line has a slope of -1, since the coefficient of pH is -1, and an intercept of the vertical axis at pH $= 0$ of $\log k_H$.

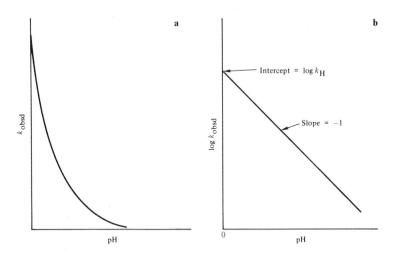

Fig. 4-2. a. Dependence of k_{obsd} on pH for a specific acid-catalyzed reaction. **b.** Dependence of $\log k_{obsd}$ on pH for the same reaction.

The hydrolysis of acetals, Eq. (18), is an example of a specific acid-catalyzed reaction.

$$(18)$$

There are three major steps in this reaction. First, hydrogen ion rapidly and reversibly protonates an oxygen of the acetal. Second, the protonated acetal fissions to yield an intermediate carbonium ion and an alcohol in a slow, rate-limiting step. Third, the carbonium ion rapidly reacts with water to yield an aldehyde and second alcohol molecule.

Specific base-catalyzed reactions which include the hydrolyses of esters and amides may be treated in a similar manner. The reaction is generalized as

$$A + OH^- \rightarrow \text{Product} + OH^-. \tag{19}$$

The rate law for this reaction is

$$-\frac{d[A]}{dt} = k_{OH}[OH^-][A], \tag{20}$$

and at any constant hydroxide ion concentration

$$k_{obsd} = k_{OH}[OH^-]. \tag{21}$$

Since hydroxide ion concentration increases with increasing pH, a plot of k_{obsd} versus pH gives an upward-curving line (Fig. 4-3a).

Since the dissociation of water to hydrogen and hydroxide ions is expressed by

$$K_w = [H^+][OH^-] = 10^{-14}, \tag{22}$$

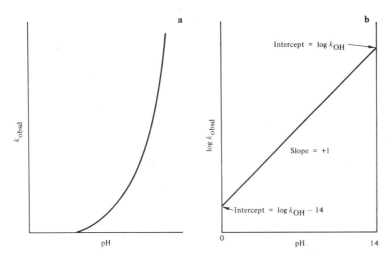

Fig. 4-3. a. Dependence of k_{obsd} on pH for a specific base-catalyzed reaction. **b.** Dependence of log k_{obsd} on pH for the same reaction.

the concentration of hydroxide ion is

$$[OH^-] = K_W/[H^+]. \tag{23}$$

Equation (23) may be substituted into Eq. (21) to obtain

$$k_{obsd} = k_{OH}K_W/[H^+]. \tag{24}$$

The log of this equation is

$$\log k_{obsd} = \log k_{OH} + \log K_W + \log(1/[H^+]), \tag{25}$$

and

$$\log k_{obsd} = \log k_{OH} - 14 + pH. \tag{26}$$

For the case of a specific base-catalyzed reaction, a plot of log k_{obsd} versus pH is a straight line (Fig. 4-3**b**). The slope of this line is $+1$ since the coefficient of pH is $+1$. The intercept of the vertical axis at pH = 0 is log $k_{OH} - 14$, and the intercept of the ordinate at pH = 14 is log k_{OH}.

An example of a specific base-catalyzed reaction is the de-aldolization of diacetone alcohol, Eq. (27).

51

$$(27)$$

In the first step of this reaction, hydroxide ion abstracts a proton from the alcohol group to produce an intermediate anion. In the slow, rate-limiting step this intermediate fissions to yield one molecule of acetone and one carbanion. The carbanion then rapidly abstracts a proton from water to yield a second molecule of acetone and regenerate hydroxide ion.

The hydrolysis of a carboxylic ester is a second example of a hydroxide ion-catalyzed reaction, Eq. (28).

$$(28)$$

In the first step of this reaction hydroxide ion attacks the electron-deficient carbonyl carbon to produce a tetrahedral intermediate anion. This intermediate may lose hydroxide ion and revert to the original ester or lose alkoxide to form the carboxylic acid. The alkoxide rapidly takes up a proton from water to produce hydroxide ion. In this reaction, hydroxide ion does not act as a base by abstracting a proton. However, since this reaction does follow the rate law of Eq. (21), it is commonly considered to be specific base catalyzed.

Chemical reactions which are dependent on the concentration of water may proceed spontaneously in aqueous solutions. Examples of such spontaneous reactions are the hydrolyses of carboxylic acid anhydrides, carboxylic acid esters that possess electron-deficient carboxyl groups (as di- and trihalo acetates), and β-lactones. Strictly speaking, water is a reactant in these reactions and not a catalyst. The generalized spontaneous reaction is,

$$A + H_2O \rightarrow Product. \tag{29}$$

The rate law followed by these reactions is,

$$-\frac{d[A]}{dt} = k_w[H_2O][A]. \tag{30}$$

Since one of the reactants, water, is in vast excess, a *pseudo* first-order reaction is observed, and

$$k_{obsd} = k_w[H_2O]. \tag{31}$$

Since k_{obsd} is independent of hydrogen ion or hydroxide ion concentration, it does not vary with pH; similarly, log k_{obsd} does not change with pH.

While reactions involving water as either a reactant or as a catalyst have rates that are independent of pH, this observation alone does not prove the participation of water; additional experimental evidence is required.

Catalysis by General Acids

A general acid catalyst is a weak acid which donates a proton in a catalytic mechanism. Weak acids undergo dissociations in pH ranges easily covered by kinetic studies. These dissociations are reflected in a dependence of rate on pH and must be taken into account in the kinetic model. A reaction catalyzed by a general acid can be represented as:

$$B^- + H^+ \underset{K_a}{\rightleftharpoons} BH$$

$$BH + A \xrightarrow{k_{BH}} Product \tag{32}$$

where BH is the general acid, B^- is its anionic form, K_a is the dissociation constant of BH, A is the reactant, and k_{BH} is the second-order rate constant for this reaction. The rate of this reaction is dependent upon the concentrations of reactant A and the general acid BH:

$$-\frac{d[A]}{dt} = k_{BH}[A][BH]. \tag{33}$$

The rate of a general acid-catalyzed reaction varies as a function of pH because the ionization of BH is dependent on pH. For Eq. (33) to be applicable at any pH, the concentration of BH must be expressed in terms of the total concentration of B, the hydrogen ion concentration, and K_a, the dissociation constant of BH. We begin with the material balance:

$$[B]_{total} = [B^-] + [BH]. \tag{34}$$

Since

$$K_a = \frac{[B^-][H^+]}{[BH]}, \tag{6}$$

$$[B^-] = \frac{K_a[BH]}{[H^+]}. \tag{35}$$

Substituting $[B^-]$ of Eq. (35) into Eq. (34)

$$[B]_{total} = \frac{K_a[BH]}{[H^+]} + [BH], \tag{36}$$

and

$$[B]_{total} = [BH]\left(\frac{K_a + [H^+]}{[H^+]}\right), \tag{37}$$

or

$$[BH] = [B]_{total}\left(\frac{[H^+]}{K_a + [H^+]}\right). \tag{38}$$

Since $-d[A]/dt = k_{BH}[A][BH]$, Eq. (33),

$$-\frac{d[A]}{dt} = k_{BH}[A][B]_{total}\left(\frac{[H^+]}{K_a + [H^+]}\right). \tag{39}$$

If the pH remains constant through the course of the reaction, the observed rate is first-order, and the rate constant for this reaction is given by

$$k_{obsd} = k_{BH}[B]_{total}\left(\frac{[H^+]}{K_a + [H^+]}\right) \tag{40}$$

where $([H^+]/(K_a + [H^+]))$ is the mole fraction of B_{total} in the form of the conjugate acid BH.

In practice, *pseudo* first-order rate constants, k_{obsd}, are determined at several concentrations of B_{total} at constant pH. A plot of k_{obsd} versus $[B]_{total}$ gives a straight line with a slope of $k_{obsd}/[B]_{total}$, the observed second-order rate constant of general acid catalysis. The intercept of the vertical axis at $[B]_{total} = 0$ corresponds to any reaction not catalyzed by B, i.e., reaction catalyzed by hydrogen ion, hydroxide ion, and water.

Equation (40) may be rearranged to give the second-order rate constant of general acid catalysis as a function of hydrogen ion concentration:

$$\frac{k_{obsd}}{[B]_{total}} = k_{BH}\left(\frac{[H^+]}{K_a + [H^+]}\right). \tag{41}$$

A graph of $k_{obsd}/[B]_{total}$ versus pH (Fig. 4-4**a**) has the sigmoid shape of a titration curve of a monobasic acid. The maximum value reached by $k_{obsd}/[B]_{total}$ at low pH is equal to k_{BH}, the second-order rate constant for the general acid-catalyzed reaction. The inflection point of the sigmoid curve is at $k_{BH}/2$, and corresponds to pH $= pK_a$. The pK_a values obtained from pH–rate profiles are apparent, and they are usually designated pK_{app}.

A plot of log $k_{obsd}/[B]_{total}$ versus pH gives the curved line of Fig. 4-4**b**. At low pH where $[H^+]$ is much greater than K_a, Eq. (41) reduces to

$$\frac{k_{obsd}}{[B]_{total}} = k_{BH}, \tag{42}$$

and

$$\log \frac{k_{obsd}}{[B]_{total}} = \log k_{BH}. \tag{43}$$

At pH values below the pK_a, the slope $= 0$. At high pH where K_a is much greater than $[H^+]$, Eq. (41) reduces to

$$\frac{k_{obsd}}{[B]_{total}} = \frac{k_{BH}}{K_a}[H^+], \tag{44}$$

55

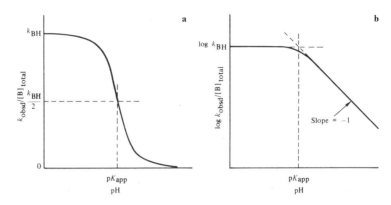

Fig. 4-4. a. Dependence of $k_{obsd}/[B]_{total}$ on pH for a general acid-catalyzed reaction. **b.** Dependence of $\log k_{obsd}/[B]_{total}$ on pH for the same reaction.

and

$$\log \frac{k_{obsd}}{[B]_{total}} = \log k_{BH} - \log K_a + \log[H^+], \qquad (45)$$

or

$$\log \frac{k_{obsd}}{[B]_{total}} = \log k_{BH} - \log K_a - pH \qquad (46)$$

Thus, at pH values above the pK_a, the slope of this plot is -1. The intersection of the two linear portions of this plot is at pH $= pK_a$.

Examples of reactions catalyzed by general acids are the hydrolyses of numerous acetals, ketals, and ortho esters. The mechanism of hydrolysis of an ortho ester is given in Eq. (47).

$$
\begin{array}{ccc}
\quad\text{R} & \text{slow,} & \quad\text{R} \\
\quad | & \text{rate} & \quad | \\
\text{BH} + \text{O}\!-\!\text{CR(OR)}_2 & \xrightarrow[\text{limiting}\ \delta\ominus]{} & \text{B}\text{-}\text{-}\text{H}\text{-}\text{-}\overset{\delta\oplus}{\text{O}}\text{-}\text{-}\text{CR(OR)}_2
\end{array}
$$

$$
\begin{array}{ccc}
\quad\text{O} & \text{fast,} & \\
\quad\parallel & +\text{H}_2\text{O} & \\
\text{BH} + \text{R}\overset{}{\text{C}}\!-\!\text{OR} & \longleftarrow & \overset{\ominus}{\text{B}} + \overset{\oplus}{\text{RC(OR)}_2} \\
\quad + \text{ROH} & & \quad + \text{ROH}
\end{array}
\qquad (47)
$$

In this case of general acid catalysis, slow proton transfer concerted with carbon–oxygen cleavage is believed to occur in the rate-limiting step.

56

Catalysis by General Bases and Nucleophiles

A general base catalyst is a weak base which abstracts a proton in a catalytic mechanism. Weak bases may also catalyze reactions by acting as nucleophiles, or Lewis bases. These two types of catalysis by weak bases are indistinguishable on the basis of their pH dependencies. Therefore, general base and nucleophilic catalysis will be considered together. The reaction catalyzed by the weak base acting as either a general base or a nucleophile is generalized by,

$$BH \overset{K_a}{\rightleftharpoons} B^- + H^+ \tag{48}$$

$$B^- + A \overset{k_B}{\rightarrow} Product$$

where B^- is the weak base, BH is its conjugate acid, K_a is the dissociation constant of BH, A is the reactant, and k_B is the second-order rate constant for this reaction. The rate of this reaction is dependent upon the concentrations of the reactant, A, and the weak base, B^-:

$$-\frac{d[A]}{dt} = k_B[A][B^-]. \tag{49}$$

The rate of this base-catalyzed reaction varies with pH because the concentration of B^- varies with pH. For Eq. (49) to be applicable at all pH values, B^- must be expressed in terms of the total concentration of B, the concentration of H^+, and the dissociation constant of BH. The material balance is used:

$$[B]_{total} = [B^-] + [BH]. \tag{50}$$

Since

$$K_a = \frac{[B^-][H^+]}{[BH]} , \tag{6}$$

$$[BH] = \frac{[B^-][H^+]}{K_a}. \tag{51}$$

Substituting [BH] of Eq. (51) into Eq. (50),

$$[B]_{total} = [B^-] + \frac{[B^-][H^+]}{K_a}, \tag{52}$$

and

$$[B]_{total} = [B^-] \left(\frac{K_a + [H^+]}{K_a} \right), \tag{53}$$

or

$$[B^-] = [B]_{total} \left(\frac{K_a}{K_a + [H^+]} \right). \tag{54}$$

Since $-d[A]/dt = k_B[A][B^-]$, Eq. (49),

$$-\frac{d[A]}{dt} = k_B[A][B]_{total} \left(\frac{K_a}{K_a + [H^+]} \right). \tag{55}$$

If the pH is kept constant through the course of the reaction, the observed rate is first-order, and the rate constant for this reaction is

$$k_{obsd} = k_B[B]_{total} \left(\frac{K_a}{K_a + [H^+]} \right), \tag{56}$$

where $(K_a/(K_a + [H^+]))$ is the mole fraction of B_{total} in the form of its conjugate base. The second-order rate constant of this weak base-catalyzed reaction, $k_{obsd}/[B]_{total}$, results if Eq. (56) is divided by $[B]_{total}$:

$$\frac{k_{obsd}}{[B]_{total}} = k_B \left(\frac{K_a}{K_a + [H^+]} \right). \tag{57}$$

Experimental values of $k_{obsd}/[B]_{total}$ are obtained as the slopes of plots of k_{obsd} versus $[B]_{total}$ at constant pH.

A plot of $k_{obsd}/[B]_{total}$ versus pH for this reaction catalyzed by a general base or a nucleophile gives a sigmoid-shaped· titration curve of a base with a single titratable group (Fig. 4-5a). The maximum value of $k_{obsd}/[B]_{total}$ is equal to k_B, and it is reached at high pH where all of B is in the form of B^-. The inflection point of the curve is at $k_B/2$, and corresponds to pH = pK_{app}.

A plot of log $k_{obsd}/[B]_{total}$ versus pH gives the curve of Fig. 4-5**b**. A logarithmic plot such as this is preferable to the nonlogarithmic plot as a means of evaluating pK_{app} if the data is limited or scattered.

At high pH where $[H^+]$ is much less than K_a, Eq. (57) reduces to

$$\frac{k_{obsd}}{[B]_{total}} = k_B, \tag{58}$$

and

$$\log \frac{k_{obsd}}{[B]_{total}} = \log k_B. \tag{59}$$

58

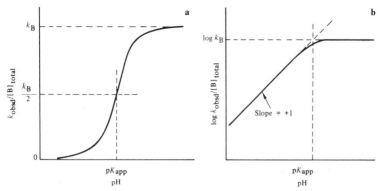

Fig. 4-5. a. Dependence of $k_{obsd}/[B]_{total}$ on pH for a general base- or nucleophile-catalyzed reaction. **b.** Dependence of log $k_{obsd}/[B]_{total}$ on pH for the same reaction.

Consequently, at pH values above the pK_a, the slope = 0. At low pH where K_a is much less than $[H^+]$, Eq. (57) reduces to

$$\frac{k_{obsd}}{[B]_{total}} = \frac{k_B K_a}{[H^+]} ,\tag{60}$$

and

$$\log \frac{k_{obsd}}{[B]_{total}} = \log k_B + \log K_a - \log[H^+],\tag{61}$$

or

$$\log \frac{k_{obsd}}{[B]_{total}} = \log k_B + \log K_a + pH.\tag{62}$$

At pH values below the pK_a, the slope of this line is $+1$. The intersection of the two linear segments of Fig. 4-5**b** occurs at pH = pK_{app}.

As stated above, a weak base acting as either a general base or a nucleophile follows the rate law of Eq. (49), and the observed second-order rate constants vary with pH as shown in Figs. 4-5**a** and 4-5**b**. These two mechanisms are indistinguishable on the basis of kinetic analysis of pH dependence. Other data are required to decide between these two possible mechanisms. A reaction involving nucleophilic catalysis may be deduced if a covalently bound intermediate can be isolated from the reaction mixture. The isolation of such an intermediate, of course, requires that its rate of formation be greater than its rate of decomposition. A mechanism of nucleophilic catalysis may also be deduced

59

if the rate of decomposition of the hypothetical covalently bound intermediate, which is synthesized independently, is identical with the rate of the overall reaction.

An example of a reaction catalyzed by a nucleophile is the hydrolysis of p-nitrophenyl acetate. The mechanism of this reaction is given in Eq. (63).

$$CH_3\overset{\overset{\displaystyle O}{\|}}{C}-O-\!\!\bigcirc\!\!-NO_2 + B^{\ominus} \rightleftharpoons CH_3\overset{\overset{\displaystyle O^{\ominus}}{|}}{\underset{\underset{\displaystyle B}{|}}{C}}-O-\!\!\bigcirc\!\!-NO_2$$

$$CH_3\overset{\overset{\displaystyle O}{\|}}{C}-OH + B^{\ominus} + H^{\oplus} \overset{H_2O}{\longleftarrow} CH_3\overset{\overset{\displaystyle O}{\|}}{C}-B + O^{\ominus}-\!\!\bigcirc\!\!-NO_2$$

(63)

In this example the nucleophile, B^-, attacks the electron-deficient carbonyl carbon to produce a tetrahedral intermediate. The intermediate decomposes to p-nitrophenolate and the acetyl derivative of the nucleophile. This derivative may then react with water to yield acetic acid and to regenerate the nucleophile.

In contrast, weak bases act as general base catalysts in the hydrolysis of ethyl acetate, Eq. (64):

$$B^{\ominus} + H_2O + CH_3\overset{\overset{\displaystyle O}{\|}}{C}-OCH_2CH_3 \rightleftharpoons CH_3\overset{\overset{\displaystyle O^{\delta\ominus}}{\|}}{C}-OCH_2CH_3$$

(64)

$$CH_3CO_2^{\ominus} + HOCH_2CH_3 \longleftarrow CH_3\overset{\overset{\displaystyle O^{\ominus}}{|}}{\underset{\underset{\displaystyle OH}{|}}{C}}-OCH_2CH_3$$

$$+ BH$$

The general base, B^-, abstracts a proton from a water molecule; simultaneously, the oxygen of the water molecule forms a bond with the carbonyl carbon of the ester. The resultant tetrahedral intermediate reverts to reactants or decomposes to products.

One may reasonably ask why a weak base will act as a nucleophilic catalyst rather than a general base for a given reaction, and vice versa. In the examples above, the answer may be found by examining the proposed tetrahedral intermediate which is produced when the nucleophile forms an adduct with the ester, Eq. (63). Viewed qualitatively, strong bases form strong chemical bonds. (For the sake of brevity and simplicity, the following treatment is qualitative; however, this explanation could also be made quantitatively in terms of ΔG's.) This fact is obvious in the example of a strong base forming a strong bond to hydrogen ion; in an analogous manner, a strong base forms a strong bond to the carbon of the tetrahedral intermediate. As the tetrahedral intermediate decomposes, the weaker bond will break and the weaker base will tend to be expelled. For example, in the reaction of p-nitrophenyl acetate, Eq. (63), the leaving group, p-nitrophenolate, is a relatively weak base, and it is easily displaced by many nucleophiles.

However, in the hydrolysis of ethyl acetate, Eq. (64), the leaving group, ethoxide ion, is a very strong base. It forms a strong bond to the carbon of the tetrahedral intermediate in the same way that it forms a strong bond to hydrogen ion. If a weak base forms a tetrahedral intermediate with ethyl acetate via nucleophilic attack, it is expelled in preference to ethoxide ion. Hence, the tetrahedral intermediate reverts back to reactant, and the nucleophilic displacement reaction is so unfavorable that it is not observed. A weak base, however, may act as a general base by assisting hydroxide to bind to the tetrahedral intermediate derived from the ester, Eq. (64). The tetrahedral intermediate thus formed may expel either of two very strong bases which are very strongly bonded to it. In the example given in Eq. (64), either ethoxide or hydroxide ions may be expelled. If ethoxide is expelled, net hydrolysis takes place. While the tetrahedral intermediate may also expel hydroxide ion and revert to reactant, all of the reactant will eventually be hydrolyzed because of the irreversibility of the final step.

Simultaneous Catalysis by a General Acid and a Weak Base

A catalytic mechanism may be envisioned in which the rate is dependent upon the concentrations of a weak acid and a weak base. This theoretical mechanism has gained attention because it would result in a "bell-shaped" dependence of observed rate constant on pH (Fig. 4-6). Bell-shaped pH dependencies occur in the reactions catalyzed by lysozyme, ribonuclease, fumarase, and numerous other enzymes. They have often been interpreted as implying the simultaneous participation of a general acid and a weak base, either as a general base or as a nucleophile, in the catalytic mechanism.

In a simple case, a bell-shaped pH dependence results if reactant A is converted to product by the monoacid form (BH^-) of a diacid (BH_2). The overall reaction is:

$$BH_2 \underset{+H^+}{\overset{\underset{-H^+}{K_{a1}}}{\rightleftharpoons}} BH^- \underset{+H^+}{\overset{\underset{-H^+}{K_{a2}}}{\rightleftharpoons}} B^{-2} \tag{65}$$

$$BH^- + A \overset{k}{\rightarrow} \text{Product.}$$

The rate of this reaction is given by,

$$-\frac{d[A]}{dt} = k[BH^-][A]. \tag{66}$$

As in the preceding examples, it is more meaningful to express the concentration of BH^- in terms of B_{total}. The material balance is

$$[B]_{total} = [BH_2] + [BH^-] + [B^{-2}]. \tag{67}$$

By employing the acid–base equilibria, one can show that

$$[B]_{total} = \frac{[BH^-][H^+]}{K_{a1}} + [BH^-] + \frac{K_{a2}[BH^-]}{[H^+]}. \tag{68}$$

Solving for $[BH^-]$,

$$[BH^-] = [B]_{total} \left(\frac{K_{a1}[H^+]}{K_{a1}K_{a2} + K_{a1}[H^+] + [H^+]^2} \right), \tag{69}$$

or

$$[BH^-] = [B]_{total} \frac{K_{a1}[H^+]}{K_{a1}(K_{a2} + [H^+]) + [H^+]^2}. \tag{70}$$

When Eq. (70) is substituted into Eq. (66), the rate of the reaction is expressed as,

$$-\frac{d[A]}{dt} = k[A][B]_{total} \frac{K_{a1}[H^+]}{K_{a1}(K_{a2} + [H^+]) + [H^+]^2}. \tag{71}$$

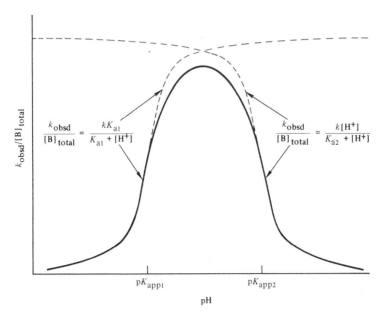

Fig. 4-6. Dependence of $k_{obsd}/[B]_{total}$ on pH for a reaction having simultaneous catalysis by a general acid and a weak base.

If the concentration of B_{total} is much greater than A, *pseudo* first-order kinetics will result, and

$$k_{obsd} = k[B]_{total} \frac{K_{a1}[H^+]}{K_{a1}(K_{a2} + [H^+]) + [H^+]^2} \,. \tag{72}$$

Equation (72) may be divided by the concentration of B_{total} to give the second-order rate constant as a function of hydrogen ion concentration:

$$\frac{k_{obsd}}{[B]_{total}} = \frac{k\,K_{a1}[H^+]}{K_{a1}(K_{a2} + [H^+]) + [H^+]^2} \,, \tag{73}$$

or

$$\frac{k_{obsd}}{[B]_{total}} = \frac{k\,K_{a1}[H^+]}{K_{a1}K_{a2} + K_{a1}[H^+] + [H^+]^2} \,. \tag{74}$$

63

As before, experimental values of $k_{obsd}/[B]_{total}$ are obtained as the slopes of plots of k_{obsd} versus $[B]_{total}$ at constant pH.

At high acid concentrations, $[H^+]$ is much greater than K_{a2}, and Eq. (73) reduces to

$$\frac{k_{obsd}}{[B]_{total}} = \frac{k\,K_{a1}}{K_{a1} + [H^+]}. \tag{75}$$

This is the equation which describes the dependence of rate constant on the ionization of a basic group, Eq. (57); it describes the ascending limb of the bell-shaped curve at low pH values (Fig. 4-6). At low acid concentrations, K_{a2} is much greater than $[H^+]$, and Eq. (74) reduces to

$$\frac{k_{obsd}}{[B]_{total}} = \frac{k\,[H^+]}{K_{a2} + [H^+]}. \tag{76}$$

This equation describes the dependence of rate constant on an acidic ionization, Eq. (41), and it gives the descending limb of the bell-shaped curve at high pH values (Fig. 4-6).

The preceding paragraphs have described the dependencies of rate constant on pH that are observed for various types of catalysis. They have been treated separately, but in practice more than one form of catalysis may be in operation. The overall rate expression for a given reaction may contain several terms for catalysis by hydrogen ion, hydroxide ion, water, weak acid, and weak base:

$$-\frac{d[A]}{dt} = (k_H[H^+] + k_{OH}[OH^-] + k_w + k_{BH}[BH] + k_B[B^-])[A]. \tag{77}$$

An additional term or terms for other forms of catalysis which have not been included in Eq. (77) may also contribute to the total reaction rate. The problems faced by an experimenter in performing a kinetic study are to identify which of these terms make a significant contribution to the total rate and to determine numerical values for these rate constants. Catalysis by weak acids and weak bases may be observed when these species are used as buffers. The second-order rate constants for these reactions at a given pH are obtained as slopes of plots of k_{obsd} versus total buffer concentration. When determined at a series of pH values, these second-order rate constants give the pH dependence of catalysis by weak acid or weak base. The intercepts of the ordinates of plots of k_{obsd} versus total buffer concentration give first-order rate constants for catalysis by the solvent species, hydrogen ion, hydroxide ion, and water. The pH dependence of these first-order rate constants may be a composite of the first three terms of Eq. (77). At any given pH, at most one or two of these terms will make a significant contribution to the observed first-order rate constant. Therefore, by examining these rate constants over selected portions of the pH range, one may evaluate the individual rate constants k_H, k_{OH}, and k_W as described above.

Linear Free Energy Relationships

Over the past half-century, chemists have expended much time and effort in devising empirical relationships which relate the structures of compounds to their reactivities. Their premise has been that similar substances react similarly, and that changes in structure produce related changes in reactivity. Empirical relationships which support this premise are termed *linear free energy relationships.*

Linear free energy relationships are quantitative correlations of rate and equilibrium constants. The rate constants for the reactions of a series of compounds participating in similar reactions can be related in a linear manner to equilibrium constants of these or similar compounds; these relations exist because both rate constants and equilibrium constants are functions of standard free energies. We have already seen (in Chapter 3) that the equilibrium constant for a reaction depends only on the difference in the standard free energies of reactants and products (i.e., $\Delta G^0 = -RT \ln K$). Also, according to the transition-state theory, the activated complex in a reaction is treated as being in equilibrium with the reactants (see Chapter 3); therefore, the rate constant depends on the free energy difference between reactants and the activated complex, $\Delta G^{\ddagger}$. Thus, rate constants and equilibrium constants may be related, since they both can be related to standard free energy changes of equilibrium processes.

An abstract discussion like this one acquires more meaning when examined in relation to concrete examples. Although numerous linear free energy relationships have been devised, we will consider only two: the Brønsted catalysis law and the Hammett equation.

a. The Brønsted Catalysis Law. One might expect intuitively that the stronger a Brønsted acid, the better it will be as a general acid catalyst. Similarly, the stronger a Brønsted base, the better catalyst it will be as a general base or nucleophile. These expectations are borne out in practice. The rate constant of an acid or base catalyzed reaction is related to the acidity or basicity of the catalyst, as measured by its pK_a, by the Brønsted catalysis law.

For a reaction catalyzed by a series of general acids the second-order rate constant increases as acidity increases (i.e., as pK_a decreases). The Brønsted catalysis law expresses this observation quantitatively as

$$\log k_{BH} = C_{BH} - \alpha pK_a \tag{78}$$

where k_{BH} is the second-order rate constant for catalysis by a general acid of a given pK_a. C_{BH} and α are constants. A plot of log k_{BH} for a reaction catalyzed by a series of general acids versus the pK_a's of the general acids gives a straight line (Fig. 4-7). The intercept of the vertical axis is C_{BH}, and the slope is $-\alpha$. The Brønsted coefficient, α, serves as a measure of the sensitivity of the reaction to catalysis by acid; it may vary from zero to one. A relatively small value for α results if the reaction is not very sensitive to acid strength (pK_a), and any acid is a good proton donor in the reaction regardless of its strength (pK_a). Conversely, a value of α which approaches unity indicates a great sensitivity to acid strength, and only the strongest acids (e.g., specific acid) can catalyze the reaction.

A reaction catalyzed by general bases may be treated in a similar manner. For this case the Brønsted catalysis law is expressed as

$$\log k_B = C_B + \beta pK_a \tag{79}$$

where k_B is the second-order rate constant for catalysis by a general base of a given pK_a. C_B and β are constants. A plot of log k_B for a reaction catalyzed by a series of general bases versus their pK_a's results in a straight line (Fig. 4-8). The intercept of the vertical axis is C_B, and the slope is β. The Brønsted coefficient β is a measure of the sensitivity of the reaction to base, and it may vary between zero and one. Like the Brønsted α coefficient, a relatively small value for β indicates a sensitivity to general catalysis. A value of β which approaches unity indicates that the reaction is observed to be catalyzed by only the strongest bases (e.g., specific base).

Reactions catalyzed by nucleophiles have also been found to obey the Brønsted catalysis law, Eq. (79). Values of β for these re-

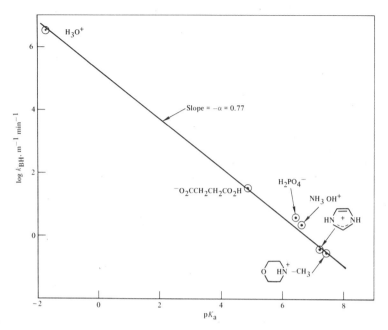

Fig. 4-7. A typical Brønsted plot for a general acid-catalyzed reaction: the general acid-catalyzed formation of the oxime of p-chlorobenzaldehyde. The general acids pertaining to each point are indicated in the figure. Data are from Jessica E. Reimann and William P. Jencks, J. Am. Chem. Soc. **88**, 3973–3982 (1966).

actions resemble those found for second-order general base-catalyzed reactions.

b. The Hammett Equation. Probably the best-known linear free energy relationship is the Hammett equation, which describes the influence of polar *meta* or *para* substituents on the side chain reactions of benzene derivatives. (It does not apply to the influence of *ortho* substituents, which may exert steric effects.) The Hammett equation relates the rate constants for reactions of these aromatic compounds to the dissociation constants of their analogously substituted benzoic acids. In examining a large number of reaction series, Hammett observed that they could be fit by the empirical, linear relationship

$$\log k = \rho \log K + C \tag{80}$$

67

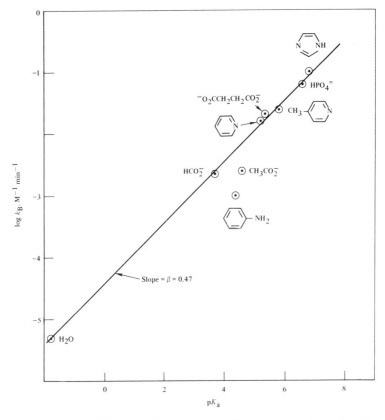

Fig. 4-8. A typical Brønsted plot for a general base-catalyzed reaction, the general base-catalyzed hydrolysis of ethyl dichloroacetate. The general bases pertaining to each point are indicated in the figure. Data are from William P. Jencks and Joan Carriuolo, *J. Am. Chem. Soc.* **83**, 1743–1750 (1961).

where k is the rate constant for reaction of a compound with a given substituent, K is the dissociation constant of the similarly substituted benzoic acid, ρ is the slope of the line, and C is constant. For the specific case where the substituent is hydrogen, Eq. (80) is

$$\log k_0 = \rho \log K_0 + C. \tag{81}$$

Here, K_0 is the dissociation constant of benzoic acid and k_0 is the rate constant for reaction of the similarly hydrogen-substituted aromatic compound. By subtracting Eq. (81) from Eq. (80), one obtains

$$\log \frac{k}{k_0} = \rho \log \frac{K}{K_0}. \tag{82}$$

Since large numbers of accurate data are available for the dissociation constants of benzoic acid, K_0, and substituted benzoic acids, K, the term $\log (K/K_0)$ can be calculated for a great number of possible substituents. This term is denoted by $\sigma = \log (K/K_0)$, and a number of σ values for aromatic substituents (calculated from data obtained in aqueous solution at 25°C) are summarized in Table 4-1. The Hammett equation, then, is commonly written as

$$\log \frac{k}{k_0} = \rho \sigma. \tag{83}$$

Table 4-1. Hammett substituent constants, σ

Substituent	meta-	para-
O^-	−0.708	−1.00
NH_2	−0.16	−0.66
CO_2^-	−0.1	0.0
CH_3	−0.069	−0.170
H	0.000	0.000
SO_3^-	+0.05	+0.09
C_6H_5	+0.06	−0.01
OCH_3	+0.115	−0.268
OH	+0.121	−0.37
SH	+0.25	+0.15
F	+0.337	+0.062
CO_2H	+0.35	+0.406
I	+0.352	+0.276
Cl	+0.373	+0.227
Br	+0.391	+0.232
CF_3	+0.43	+0.54
CN	+0.56	+0.660
NO_2	+0.710	+0.778

Source: Jack Hine, *Physical Organic Chemistry,* 2nd edition, McGraw-Hill, New York, 1962, p. 87.

A typical Hammett plot, that for the alkaline hydrolysis of ethyl benzoates, is shown in Fig. 4-9.

As noted above, σ is the *substituent constant.* The *reaction constant* is ρ, and it depends on the nature of the reaction and measures the susceptibility of the reaction to polar effects. If the value of ρ is positive, the rate of the chemical reaction is enhanced by the presence of electron-withdrawing substituents, as is the

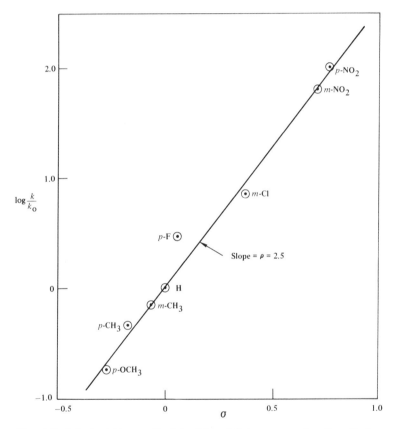

Fig. 4-9. A typical Hammett plot of log (k/k_0) versus σ for the alkaline hydrolysis of substituted ethyl benzoates in 85% ethanol at 30°C. Data are from John D. Roberts and Marjorie C. Caserio, *Basic Principles of Organic Chemistry,* Benjamin, New York, 1965, pp. 955–957.

case for the alkaline hydrolysis of ethyl benzoate. If the value of ρ is negative the rate of the reaction is increased by the presence of electron-releasing substituents.

Lastly, it should be noted that often a point will deviate from the line of a Hammett plot. These deviations may be the result of steric effects of the substituent (which makes the Hammett relationship invalid for *ortho*-substituted compounds), or strong resonance interactions between the substituent and the reaction site.

Kinetically Equivalent Rate Laws

In this treatise the study of kinetics has been approached in two major steps: (1) the collection of data and (2) the formulation of a theory which is in agreement with the data. In the examples presented above it might appear that the kinetic rate law can be deduced unambiguously from the experimental data. Unfortunately, this is not true. More than one rate law may be in agreement with the experimental data. The possibility of postulating kinetically equivalent rate laws occurs when the reactant or reactants exist in equilibrium with another form of the reactant. A closer examination of the models for "apparent general acid catalysis" and "apparent general base catalysis" illustrate this point.

A reaction which is apparently catalyzed by a general acid is (Path *a*):

$$BH + A \overset{k_{BH}}{\rightarrow} Product + BH. \tag{84}$$

The rate law for true general acid catalysis, Path *a*, Eq. (84) has already been demonstrated to be,

$$-\frac{d[A]}{dt} = k_{BH}[B]_{total} \left(\frac{[H^+]}{K_a + [H^+]} \right) [A]. \tag{39}$$

The data which lead to the deduction of apparent catalysis by a general acid may also be explained by a reaction mechanism which involves catalysis by the products of ionization of the weak acid, i.e., the conjugate base, and hydrogen ion (Path *b*):

$$BH \overset{K_a}{\rightleftharpoons} B^- + H^+$$

$$B^- + H^+ + A \overset{k_B}{\rightarrow} Product + B^- + H^+. \tag{85}$$

71

The rate law for reaction via Path *b,* Eq. (85), is:

$$-\frac{d[A]}{dt} = k_B[B^-][H^+][A]. \tag{86}$$

We have already seen that for the dissociation of BH,

$$[B^-] = [B]_{total} \left(\frac{K_a}{K_a + [H^+]} \right). \tag{54}$$

By substituting [B$^-$] of Eq. (54) into Eq. (86), one can show that

$$-\frac{d[A]}{dt} = k_B[B]_{total} \left(\frac{K_a}{K_a + [H^+]} \right)[H^+][A]. \tag{87}$$

The similarity of the forms of Eqs. (39) and (87), which describe the rates of reactions via Path *a* and Path *b,* respectively, is obvious. By equating these two expressions, one can demonstrate that

$$k_{BH} = k_B K_a. \tag{88}$$

Thus, apparent general acid catalysis, as observed in experimental data, may be the result of true general acid catalysis, Path *a* of Eq. (84). Alternatively, it may represent catalysis by the products of the ionization of the weak acid, i.e., the conjugate base and hydrogen ion, Path *b* in Eq. (85). The rate laws which describe these two paths, Eqs. (39) and (87), respectively, are kinetically equivalent; and both agree with the experimental data.

A reaction which is apparently catalyzed by a general base is (Path *a*):

$$BH \overset{K_a}{\rightleftharpoons} B^- + H^+ \tag{89}$$

$$B^- + A \overset{k_B}{\rightarrow} Product + B^-.$$

We have already seen that the rate law for true general base catalysis, Path *a,* Eq. (89), is

$$-\frac{d[A]}{dt} = k_B[B]_{total} \left(\frac{K_a}{K_a + [H^+]} \right)[A]. \tag{55}$$

Apparent general base catalysis can also be explained as the result of catalysis by the conjugate general acid and hydroxide ion (Path *b*):

$$BH + OH^- + A \overset{k_{BH}}{\rightarrow} Product + BH + OH^-. \tag{90}$$

The rate law for this path is

$$-\frac{d[A]}{dt} = k_{BH}[BH][OH^-][A]. \tag{91}$$

We have already demonstrated that

$$[BH] = [B]_{total} \left(\frac{[H^+]}{K_a + [H^+]} \right). \tag{38}$$

Therefore,

$$-\frac{d[A]}{dt} = k_{BH}[B]_{total} \left(\frac{[H^+]}{K_a + [H^+]} \right) [OH^-][A]. \tag{92}$$

Since $[H^+][OH^-] = K_w$, Eq. (92) becomes,

$$-\frac{d[A]}{dt} = k_{BH}[B]_{total} \left(\frac{K_w}{K_a + [H^+]} \right) [A]. \tag{93}$$

The similarity of the rate laws for Paths a and b of Eqs. (89) and (92) as given by Eqs. (55) and (93), respectively, is obvious. If these two expressions are equated, one can show that

$$k_B K_a = k_{BH} K_w. \tag{94}$$

Thus, in the case of apparent general base catalysis, more than one reaction pathway can be postulated, and more than one rate law can be derived which is in agreement with the data.

Many other examples could be given in which kinetically equivalent rate laws may be derived to fit a set of kinetic data. A major problem of kinetic analysis is the definition of all possible kinetically equivalent rate laws and determination of which of these actually describes the reaction under consideration. Such determinations are left for discussion in the context of specific examples.

While a kinetic rate law may be expressed as a function of the concentration of the reactants, transition-state theory holds that the kinetic rate law is a function of the concentration of activated complex (Chapter 3). The salient point here is that kinetics does not give the stoichiometry of the reactants; kinetics gives the stoichiometry of the transition state. In many reactions the concentration of activated complex may be expressed simply as a function of the concentrations of reactants. If, however, one or more of the reactants exists in an equilibrium with a related form, the form which reacts may not be deduced from the kinetics. Stated more concretely, if one or more of the reactants undergoes an ionic equilibrium, as in the examples above, more than one rate law may be written which is in agreement with the experimental data. Analysis of the kinetic data cannot reveal where or how a proton is involved in a transition state. Stated generally, kinetic analysis gives the stoichiometry of the transition state; it does not identify the reacting species. This rule-of-thumb should be considered whenever one attempts to write a rate law to fit experimental data.

REFERENCES

Myron L. Bender, *Mechanisms of Homogeneous Catalysis from Protons to Proteins,* Part One, Wiley-Interscience, New York, 1971.

Thomas C. Bruice and Stephen Benkovic, *Bioorganic Mechanisms,* Vol. 1, Chapter 1, Benjamin, New York, 1966.

William P. Jencks, *Catalysis in Chemistry and Enzymology,* Chapters 3 and 11, McGraw-Hill, New York, 1969.

Eugene Zeffren and Philip L. Hall, *The Study of Enzyme Mechanisms,* Chapter 6, Wiley-Interscience, New York, 1973.

Jack Hine, *Physical Organic Chemistry,* 2nd edition, Chapters 4 and 5, McGraw-Hill, New York, 1969.

PROBLEMS

1. For a reaction that follows the rate law for general acid catalysis, demonstrate that at $k_{obsd}/[B]_{total} = k_{BH}/2$, pH = pK_a. You may use as a starting point any equation in the text.

2. For the same reaction, show that in the log $k_{obsd}/[B]_{total}$ versus pH profile the two linear segments intersect at pH = pK_{app}. Again, any equation in the text is a valid starting point.

3. Edwards studied the hydrolysis of aspirin over a wide pH range (L. J. Edwards, *Trans. Faraday Soc.* **47**, 723 (1950)). His observed *pseudo* first-order rate constants at varied pH values are in the accompanying table.

pH	k_{obsd} (days^{-1})
0.53	0.578
1.00	0.190
1.33	0.0835
1.80	0.0450
2.48	0.0267
2.99	0.0343
3.53	0.0561
4.04	0.0880
4.49	0.125
5.03	0.130
5.64	0.130
6.04	0.120
6.06	0.120
6.21	0.113
6.64	0.110
6.98	0.107
7.40	0.108
7.75	0.116
8.10	0.134
7.35	0.115
7.37	0.120
8.10	0.134
8.19	0.154
9.48	0.321
10.52	1.97
11.29	13.7
12.00	58.0
12.21	121.0
12.48	273.0
12.66	476.0
12.77	530.0

By following the procedures outlined in the preceding pages, determine by graphical methods the four rate constants of the experimental rate law:

$$k_{obsd} = k_H[H^+] \left(\frac{[H^+]}{K_a + [H^+]} \right) + k_{H_2O} \left(\frac{[H^+]}{K_a + [H^+]} \right) +$$
$$k'_{H_2O} \left(\frac{K_a}{K_a + [H^+]} \right) + k_{OH}[OH^-] \left(\frac{K_a}{K_a + [H^+]} \right).$$

The pK_a for the carboxyl group of aspirin is 3.57.

4. The hydrolysis of ethyl trifluorothiolacetate is catalyzed by phosphate buffers (L. Fedor and T. C. Bruice, *J. Am. Chem. Soc.* **87**,

4138 (1965)). Second-order rate constants for this reaction determined at several pH values are given in the accompanying table:

pH	$k_{obsd}/[B]_{total}(M^{-1} min^{-1})$
5.4	0.5
5.9	1.0
6.2	1.4
6.7	2.4
7.0	3.3
7.6	4.8
8.0	5.2
8.5	5.4

a. Write as many kinetically indistinguishable rate laws which fit these data as you can.

b. What is the pK_{app} of the phosphate catalyst?

5. The mixed anhydride acetyl phosphate hydrolyzes at a measurable rate over the entire pH range (G. DiSabato and W. P. Jencks, *J. Am. Chem. Soc.* **83**, 4400 (1961)). The observed *pseudo* first-order rate constants for this reaction at varied pH values are as in the table.

pH	$k_{obsd} \times 10^{-3}(min^{-1})$
0.2	65.0
0.5	40.0
1.2	20.0
2.0	11.0
3.0	10.0
4.0	9.0
4.5	7.0
5.0	5.0
6.0	4.0
7.0	4.0
8.0	4.0
9.0	4.0
10.0	4.5
11.0	9.0
11.5	25.0
12.0	80.0

Write one (the simplest) rate law which fits this data. Evaluate the rate constant(s) and equilibrium constant involved. Assume that only the mono and di-ionized form of the reactant exist in the pH range under study.

6. Compound (I) has been found to hydrolyze with the pH dependence shown in the figure (T. C. Bruice and D. Piszkiewicz, *J. Am. Chem. Soc.* **89**, 3568 (1967)).

a. Write two rate laws that are consistent with this log k_{obsd} versus pH profile.

b. Show or briefly describe the mechanism or mechanisms which are in agreement with the terms of these rate laws which predominate at pH values greater than 4.

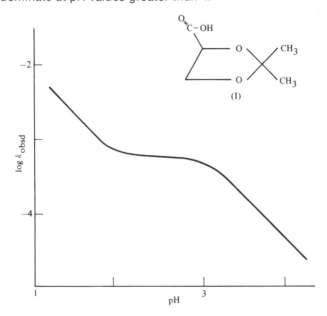

7. Weak bases react with *p*-nitrophenyl acetate by a nucleophilic displacement mechanism which liberates *p*-nitrophenolate. Second-order rate constants for the reaction of several bases with *p*-nitrophenyl acetate and their pK_a values are as in the accompanying table (T. C. Bruice and R. Lapinski, *J. Am. Chem. Soc.* **80**, 2265–2267 (1958)).

Nucleophile	pK_a	k_2(M^{-1} min^{-1})
$C_6H_5O^-$	9.9	19.0
p-$ClC_6H_4O^-$	9.3	9.6
p-CHO-$C_6H_4O^-$	7.6	0.5
HPO_4^{-2}	7.0	0.0151
$CH_3CO_2^-$	4.7	0.00095

a. What is the Brønsted β coefficient for this reaction?

b. Can you estimate the rate constant for the reaction of bicarbonate (HCO_3^-), which has a pK_a of 6.35, with p-nitrophenyl acetate?

8. The pK_a values of *meta*- and *para*-trimethylaminobenzoic acids are 3.45 and 3.43, respectively; that of benzoic acid is 4.20. What are the values of σ for these substituents?

9. Aldehydes react with the amino compound semicarbazide to form semicarbazones:

$$R-CHO + H_2N-NH\overset{\displaystyle O}{\overset{\|}{C}}NH_2 \rightarrow R-CH=N-NH\overset{\displaystyle O}{\overset{\|}{C}}NH_2.$$

Second-order rate constants for this reaction with several substituted benzaldehydes are as follows(B. M. Anderson and W. P. Jencks, *J. Am. Chem. Soc.* **82**, 1773–1777 (1960)):

Substituent	k_{obsd} at pH 1.75 in 0.02 M semicarbazide (min^{-1})
p-OCH$_3$	0.157
p-CH$_3$	0.195
H	0.304
p-Cl	0.347
m-NO$_2$	1.02
p-NO$_2$	1.50

a. What is the value of ρ for this reaction?

b. What conclusions regarding the susceptibility of this reaction to the electronic effects of substituents can you draw from these data?

3

ENZYME
CATALYSIS
AND KINETICS

The simplest reaction catalyzed by an enzyme is described by the equation,

$$E + S \rightarrow E + Product, \tag{1}$$

where E is the enzyme and S is the reactant, or substrate, being converted to product. Since the enzyme is a catalyst it is unchanged by the reaction, and its total concentration remains constant during the reaction.

The kinetic form, or kinetic order, taken by an enzyme-catalyzed reaction may vary considerably with the conditions used. Let us consider, for example, a series of kinetic experiments in which the enzyme concentration is held constant, but the initial substrate concentraton is varied. If the initial substrate concentration is low, the loss of substrate in the overall reaction, $-d[S]/dt$, may appear to be a first-order process (Fig. 5-1**a**). If the initial substrate concentration is very high, the rate of loss of substrate, $-d[S]/dt$, may appear to remain constant over a substantial portion of the reaction (Fig. 5-1**c**). In this case, the rate of the reaction is independent of substrate concentration, that is

$$-\frac{d[S]}{dt} = k[E]. \tag{2}$$

Such a reaction is zero-order, since the exponent of the substrate concentration in the rate expression is zero (i.e., $[S]^0 = 1$, and is not shown in the equation). At intermediate initial substrate concentrations, the rate of conversion of substrate to product may proceed by a complex reaction which is neither first-order nor zero-order (Fig. 5-1**b**). The rate law followed by an enzyme-catalyzed reaction under a given set of conditions is not obvious at the outset, and it may be very difficult to evaluate.

To avoid these possible problems, studies of enzyme-catalyzed reactions are generally conducted by measuring their initial rates or velocities. Plots of substrate concentration versus time (Fig. 5-1**a**, **b**, **c**) generally give straight lines for the first 10 to 20% of the total reaction. Their slopes are equal to the initial reaction rates or velocities, $-d[S]/dt$, which are described by the term v. The initial velocities of enzyme-catalyzed reactions, v, are zero-order rate constants with units of concentration $\times$ time^{-1}.

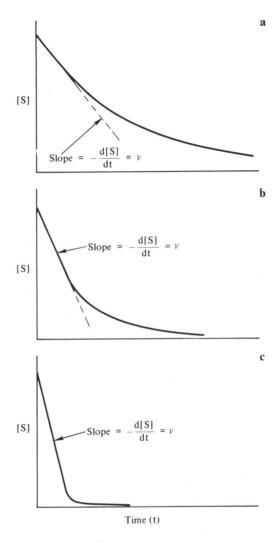

Fig. 5-1. Progress curves for enzyme-catalyzed reactions at three different initial substrate concentrations. **a.** Very low initial substrate concentration may result in an apparently first-order process. **b.** Intermediate initial substrate concentration may result in a reaction of complex kinetic order. **c.** Very high initial substrate concentration may result in a zero-order reaction.

The Enzyme–Substrate Complex

The rates of enzyme-catalyzed reactions show a characteristic dependence on substrate concentration. A plot of the initial velocities, v, versus substrate concentrations yields a curved line (Fig. 5-2). At low substrate concentration the initial velocity is proportional to the product of enzyme and substrate concentrations (i.e., $v = k[E][S]$). At high substrate concentrations the velocity is proportional to the concentration of enzyme alone (i.e., $v = k[E]$). This diphasic dependence of rate on substrate concentration was first observed independently by Brown and Henri in 1902, and it led to the proposal of the formation of an intermediate complex between enzyme and substrate. In 1913, Michaelis and Menten developed the first satisfactory mathematical treatment of the effect of substrate concentration on the velocity of an enzyme-catalyzed reaction which incorporated the concept of the formation of a complex of enzyme and substrate.

The Michaelis-Menten treatment assumes the reversible formation of an enzyme–substrate intermediate, which is followed by a first-order decomposition of complex to product. (The derivation presented below is that of Haldane. While it reaches the same

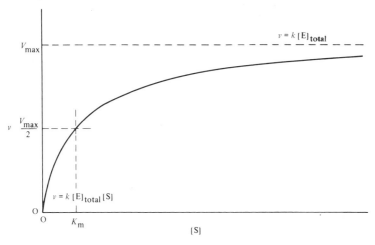

Fig. 5-2. The effect of substrate concentration on the velocity of an enzyme-catalyzed reaction.

conclusion as the original Michaelis-Menten treatment, it is preferable since it requires fewer assumptions.) The overall reaction is written as

$$E + S \underset{k_2}{\overset{k_1}{\rightleftharpoons}} ES \overset{k_3}{\rightarrow} E + Product, \tag{3}$$

where E is free enzyme, S is substrate, and ES is the enzyme–substrate complex. The overall velocity of this reaction is expressed by

$$v = k_3[ES]. \tag{4}$$

In deriving an equation for the velocity, v, as a function of [S], [S] must be expressed in terms of [ES]. This expression for [ES] may then be substituted into Eq. (4). In this derivation the following terms are used:

$[E]_{total}$ = total concentration of enzyme

[S] = total concentration of substrate, such that [S] is much greater than $[E]_{total}$

[ES] = concentration of enzyme–substrate complex

$[E]_{total}$ − [ES] = concentration of free enzyme.

To begin, the rate of formation of ES is given by

$$-\frac{d[ES]}{dt} = k_1([E]_{total} - [ES])[S]. \tag{5}$$

The rate of decomposition of ES back to E and S, and to products is

$$-\frac{d[ES]}{dt} = k_2[ES] + k_3[ES]. \tag{6}$$

During the period of reaction where initial velocities are measured, the concentrations of substrate, S, free enzyme, and total enzyme, will remain substantially constant. It may be legitimately assumed that the concentration of the enzyme–substrate complex, [ES], will also be constant, or in a "steady state." If this steady-state assumption is made, then the rate of formation of ES is equal to the rate of its decomposition:

$$k_1([E]_{total} - [ES])[S] = k_2[ES] + k_3[ES]. \tag{7}$$

This may be rearranged to isolate the rate constants:

$$\frac{([E]_{total} - [ES])[S]}{[ES]} = \frac{k_2 + k_3}{k_1} = K_m. \tag{8}$$

The isolated rate constants, $(k_2 + k_3)/k_1$, are replaced by the Michaelis-Menten constant, K_m, which has the units of concentration. The steady-state concentration of ES may be found by solving Eq. (8) for ES:

$$[ES] = \frac{[E]_{total}[S]}{K_m + [S]} . \tag{9}$$

The concentration of ES as given in Eq. (9) may now be substituted into Eq. (4) to give the velocity of the enzyme-catalyzed reaction as a function of substrate concentration:

$$v = \frac{k_3[E]_{total}[S]}{K_m + [S]} . \tag{10}$$

Since the total concentration of enzyme, $[E]_{total}$, is not always known, this term is replaced. According to the Michaelis-Menten model, at very high substrate concentrations, essentially all of the enzyme is complexed with substrate (i.e., $[E]_{total} = [ES]$). Since $v = k_3 [ES]$, Eq. (4), a maximum velocity, V_{max}, is obtained when

$$v = V_{max} = k_3[E]_{total}. \tag{11}$$

The Michaelis-Menten equation is obtained by substituting V_{max} for $k_3[E]_{total}$ in Eq. (10):

$$v = \frac{V_{max}[S]}{K_m + [S]} . \tag{12}$$

This equation quantitatively relates the dependence of initial velocity, v, to the concentration of substrate as seen in Fig. 5-2. At low substrate concentration where [S] is much less than K_m Eq. (12) reduces to $v = (V_{max}/K_m)[S]$. Under this condition velocity is proportional to substrate concentration and the term V_{max}/K_m is an apparent first-order rate constant for the enzymatic reaction. Also at high substrate concentration where [S] is much greater than K_m, Eq. (12) reduces to $v = V_{max} = k_3[E]_{total}$, and the reaction velocity is independent of substrate concentration.

In Eq. (12), the substrate concentration and K_m have the same units: concentration. If we set K_m equal to [S], the Michaelis-Menten equation becomes,

$$v = \frac{V_{max}K_m}{K_m + K_m} , \tag{13}$$

and

$$v = \frac{V_{max}}{2}. \tag{14}$$

Thus, when the observed initial velocity, v, is half of the maximum obtainable velocity, V_{max}, the K_m is equal to the substrate concentration and may be evaluated from a plot such as Fig. 5-2.

As noted above, the Michaelis-Menten constant, K_m, is equal to a composite of rate constants:

$$K_m = \frac{k_2 + k_3}{k_1}. \tag{15}$$

In some enzyme-catalyzed reactions, k_2 and k_1 are very large relative to k_3. When this is the case, k_3 in Eq. (3) is much smaller than k_2, and Eq. (15) simplifies to

$$K_m \approx \frac{k_2}{k_1}. \tag{16}$$

In this special case K_m is approximately equal to the dissociation constant for the ES complex. The actual dissociation constant of the ES complex is termed K_S. Unfortunately, K_m is often indiscriminately confused with K_S. The assumption that these two terms are approximately identical must not be made unless specific experimental information indicates that k_3 is very much smaller than k_2.

The plot of observed initial velocity, v, versus substrate concentration (Fig. 5-2) illustrates a property of enzyme-catalyzed reactions: saturation of enzyme with substrate results in a maximum obtainable velocity. The difficulty in experimentally obtaining the value of the maximum velocity and the curved nature of this plot make other methods of deriving K_m and V_{max} desirable.

Three linear transformations of the Michaelis-Menten equation have been used most commonly to obtain K_m and V_{max}. The most commonly used, the method of Lineweaver and Burk, takes the reciprocal of Eq. (12):

$$\frac{1}{v} = \frac{K_m + [S]}{V_{max}[S]}, \tag{17}$$

or

$$\frac{1}{v} = \frac{K_m}{V_{max}} \left(\frac{1}{[S]} \right) + \frac{1}{V_{max}}. \tag{18}$$

A plot of $1/v$ versus $1/[S]$ (Fig. 5-3) yields a straight line which has a slope of K_m/V_{max}. The intercept of the vertical axis is $1/V_{max}$; the intercept of the horizontal axis is $-1/K_m$. This method has the advantage of being conceptually easy to grasp in that it keeps the variables of velocity and concentration separate. In addition, it is useful in studies of enzyme inhibition (discussed below).

The second linear transformation of the Michaelis-Menten equation, which was also derived by Lineweaver and Burk, multiplies Eq. (18) by $[S]$ to obtain

$$\frac{[S]}{v} = \frac{K_m}{V_{max}} + \frac{[S]}{1}\frac{1}{V_{max}}. \tag{19}$$

According to this transformation, a plot of $[S]/v$ versus $[S]$ (Fig. 5-4) will yield a straight line with a slope of $1/V_{max}$ and an intercept of the vertical axis of K_m/V_{max}. The intercept of the horizontal axis is $-K_m$.

The third linear transformation of the Michaelis-Menten equation, which is often used, is the Eadie-Hofstee plot. Equation (18) may be multiplied by $V_{max}v$ to obtain

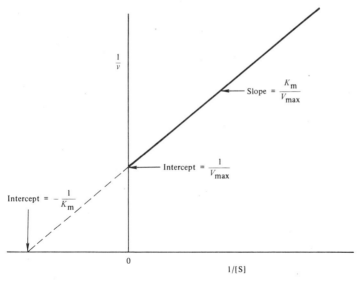

Fig. 5-3. Plot of $1/v$ versus $1/[S]$ according to the method of Lineweaver and Burk.

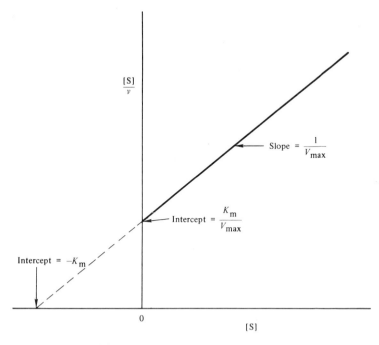

Fig. 5-4. Plot of [S]/v versus [S] according to the method of Lineweaver and Burk.

$$V_{max} = K_m \frac{v}{[S]} + v, \tag{20}$$

or

$$v = V_{max} - K_m \frac{v}{[S]}. \tag{21}$$

A plot of v versus $v/[S]$ also yields a straight line (Fig. 5-5) which has a slope of $-K_m$ and an intercept of the vertical axis of V_{max}. The intercept of the horizontal axis is V_{max}/K_m. The Eadie-Hofstee plot is used often because it gives values of V_{max} and K_m directly, and it exaggerates deviations from linearity, which is predicted by the Michaelis-Menten model.

The Michaelis-Menten treatment provided a mathematically acceptable and esthetically pleasing formulation describing enzyme-catalyzed reaction velocities based on the assumption of enzyme–

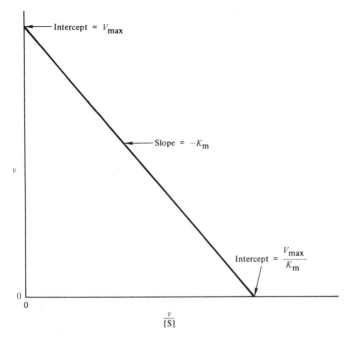

Intercept = V_{max}

Slope = $-K_m$

v

Intercept = $\dfrac{V_{max}}{K_m}$

0

0

$\dfrac{v}{[S]}$

Fig. 5-5. Plot of v versus $v/[S]$ according to the method of Eadie and Hofstee.

substrate complex formation. However, it did not prove that an enzyme–substrate intermediate is formed in an enzyme-catalyzed reaction. Direct experimental evidence for the existence of such an intermediate was required. In 1943 Chance provided this evidence by studying spectroscopically the reaction of hydrogen peroxide with peroxidase. He found that this reaction resulted in a pronounced perturbation of the absorption spectrum of the heme group of the enzyme, thus demonstrating the formation of an enzyme–substrate complex.

Further inescapable proof of the formation of the enzyme–substrate intermediate has been furnished by the identification of covalently bound enzyme–substrate compounds. In a kinetic study of the hydrolysis of p-nitrophenyl acetate by chymotrypsin, Hartley and Kilby presented the first evidence which suggested the formation of a covalently bound intermediate. The progress of this reaction was followed with a spectrophotometer, by measuring the formation of the yellow *p*-nitrophenolate product from the colorless reactant. The time course of the reaction (Fig. 5-6) showed a rapid burst in the formation of product in a pre-steady-state reaction. After the rapid burst, the alcohol was produced at a slower, constant velocity. When the line describing the slow, constant rate of alcohol production was extrapolated back to zero time, it intersected the vertical axis at a concentration of *p*-nitrophenolate equal to the concentration of chymotrypsin used in the experiment. Quantitatively, this product corresponded to that amount produced in the initial burst, not in the slow steady-state reaction.

The mechanism of the hydrolysis of *p*-nitrophenyl acetate and other substrates by chymotrypsin deduced from this kinetic data is a two-step process. In the first step, after formation of the enzyme–substrate complex, one molecule of substrate reacts with one molecule of enzyme to yield one molecule of alcohol and a covalent acyl–enzyme intermediate. The production of alcohol in this first step corresponds to the rapid burst observed in the kinetic experiment (Fig. 5-6). In a second rate-limiting step, the acyl–enzyme intermediate decomposes slowly to free enzyme and carboxylic acid. The overall reaction, which includes the rapid

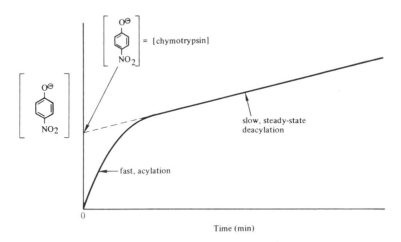

Fig. 5-6. Chymotrypsin-catalyzed production of *p*-nitrophenolate from *p*-nitrophenyl acetate as a function of time.

formation of a noncovalently bound enzyme–substrate complex, is written as,

$$Ac{-}OR + E \xrightleftharpoons{} Ac{-}OR \cdot E \xrightarrow{\text{fast}} Ac{-}OE \xrightarrow[H_2O]{\text{slow}} Ac + E. \qquad (22)$$
$$+$$
$$ROH$$

Chymotrypsin reacts with p-nitrophenyl trimethylacetate to yield an unusually stable trimethylacetyl–enzyme intermediate [Ac—OE in Eq. (22)]. This stability allowed McDonald and Balls to isolate and crystallize trimethylacetyl chymotrypsin. Similar stable enzyme–substrate compounds have been isolated for a number of other cases. Thus, enzymes not only form noncovalent complexes with their substrates, but they may also form covalently bound intermediates as part of their catalytic mechanisms.

Inhibition of Enzyme Catalysis

An inhibitor is a substance which, when it interacts with an enzyme, causes a decrease in enzyme catalytic activity. There are naturally occurring inhibitors of enzymes which bring about a control of metabolism within the cell. Synthetic inhibitors of enzymes are also of great importance; these include drugs and poisons, e.g., insecticides. Inhibitors are used as tools in the study of metabolism within the cell. For example, an inhibitor may be used to inactivate one enzyme of a pathway, thereby allowing substrate and precursors to accumulate in sufficient quantities for isolation and identification. Inhibitors are also used to probe the mechanisms of enzymatic reactions. They are powerful tools in ascertaining the order of addition of substrates and release of products in multiple-substrate enzymatic reactions. (This use of inhibitors is discussed in Chapter 6.)

Enzyme inhibition may be reversible or irreversible. It is generally accepted that reversible inhibition results from noncovalent complex formation between enzyme and inhibitor. Irreversible inhibition is usually found to result from covalent bond formation between enzyme and inhibitor. Only reversible inhibition is discussed here. Three simple types of inhibition are recognized. They are termed *competitive, noncompetitive,* and *uncompetitive.* In addition, inhibition of an enzymatic reaction by an excess of its own substrate has been observed. These types of inhibition are generally distinguished by the effect they have on the Lineweaver-Burk plots of the enzyme-catalyzed reactions.

a. Competitive Inhibition.   The model for competitive inhibition assumes that substrate and inhibitor compete for the free enzyme; i.e., complexation of substrate, S, and inhibitor, I, to the enzyme, E, are mutually exclusive.   The reactions involved are expressed by,

$$E + S \rightleftharpoons ES \rightarrow E + Product, \tag{23}$$
$$K_i \Updownarrow + I$$
$$EI$$

where K_i is the dissociation constant of the enzyme–inhibitor complex, EI.   The addition of this equilibrium makes the velocity of the reaction dependent upon the concentration of inhibitor, I, and K_i as well as the concentration of substrate, S, and K_m.   The Michaelis-Menten equation for the model of competitive inhibition becomes,

$$v = \frac{V_{max}[S]}{K_m \left(1 + \dfrac{[I]}{K_i}\right) + [S]}. \tag{24}$$

The presence of inhibitor causes K_m to be multiplied by the factor $(1 + [I]/K_i)$.   The Lineweaver-Burk transformation becomes,

$$\frac{1}{v} = \frac{K_m}{V_{max}} \left(\frac{1}{[S]}\right)\left(1 + \frac{[I]}{K_i}\right) + \frac{1}{V_{max}}. \tag{25}$$

According to Eq. (25), a plot of $1/v$ versus $1/[S]$ gives a straight line with a slope of $(K_m/V_{max})(1 + [I]/K_i)$ (Fig. 5-7).   The vertical intercept at $1/[S] = 0$ is $1/V_{max}$, and the intercept of the horizontal axis is $-1/K_m(1 + [I]/K_i)$.   Thus, the presence of a constant concentration of competitive inhibitor increases the slope of the Lineweaver-Burk plot (Fig. 5-7), but has no effect on the observed value of V_{max}.

An example of competitive inhibition is the reaction catalyzed by succinate dehydrogenase.   This enzyme catalyzes the FAD-dependent oxidation of succinate to fumarate:

$$
\begin{array}{l}
CO_2^{\ominus} \qquad\quad {}^{\ominus}O_2C \\
| \qquad\qquad\qquad \diagdown \\
CH_2 \qquad\qquad\qquad CH \\
| \quad + \;\; E{-}FAD \rightarrow \quad || \qquad + \; E{-}FADH_2. \\
CH_2 \qquad\qquad\qquad HC \\
| \qquad\qquad\qquad \diagup \\
CO_2^{\ominus} \qquad\qquad\quad CO_2^{\ominus}
\end{array}
\tag{26}
$$

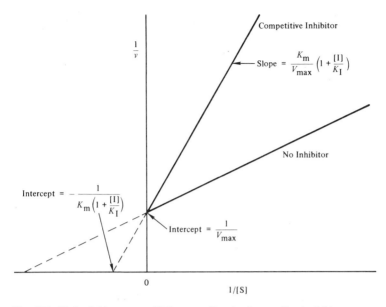

Fig. 5-7. Plot of $1/v$ versus $1/[S]$ according to the method of Lineweaver and Burk. Experiments carried out in the absence and in the presence of a competitive inhibitor.

Substances whose presence causes the kinetics of this reaction to fit the model of competitive inhibition include oxalate and malonate. These inhibitors are di-anionic compounds like the substrate of the enzyme, succinate. The similarities in structure and the mutually exclusive complexation of substrate or inhibitor with enzyme, which is deduced from the theoretical model, suggest that substrate and inhibitors bind to the same site on the enzyme surface.

    b. Noncompetitive Inhibition. A second type of inhibition which is identified on the basis of kinetic analysis is noncompetitive inhibition. It has been seen in single substrate reactions, but is far more common in multiple substrate systems. The classical model for noncompetitive inhibition of a single substrate reaction assumes the inhibitor, I, may complex both the free enzyme, E, and the enzyme–substrate complex, ES. All of the reactions taking place are summarized by

$$E \underset{-S}{\overset{+S}{\rightleftharpoons}} ES \rightarrow E + Product.$$

$$K_i \updownarrow +I \qquad K_i \updownarrow +I$$

$$EI \underset{-S}{\overset{+S}{\rightleftharpoons}} ESI \qquad (27)$$

For the sake of simplicity, it is assumed that K_i is the same for dissociation of inhibitor from EI and from ESI. The Michaelis-Menten equation which describes the velocity of this reaction is

$$v = \frac{V_{max}[S]/\left(1 + \frac{[I]}{K_i}\right)}{K_m + [S]} . \qquad (28)$$

The Lineweaver-Burk transformation of this equation is

$$\frac{1}{v} = \left[\frac{K_m}{V_{max}}\left(\frac{1}{[S]}\right) + \frac{1}{V_{max}}\right]\left(1 + \frac{[I]}{K_i}\right). \qquad (29)$$

A plot of $1/v$ versus $1/[S]$, according to this equation, gives a straight line which has a slope of $(K_m/V_{max})(1 + [I]/K_i)$ (Fig. 5-8). The vertical intercept at $1/[S] = 0$ is $(1/V_{max})(1 + [I]/K_i)$, and the horizontal intercept at $1/v = 0$ is $-1/K_m$.

In the theoretical model, Eq. (27), the assumption was made that the noncompetitive inhibitor could combine with both the free enzyme and the enzyme–substrate complex. For this to take place, the inhibitor must bind at a site other than that to which substrate binds.

c. Uncompetitive Inhibition. A third type of inhibition is uncompetitive inhibition. Like noncompetitive inhibition, it is rare in single substrate reactions, but common in multiple substrate systems. The classical model for uncompetitive inhibition of a single substrate reaction assumes that inhibitor, I, may combine only with the enzyme–substrate complex. The reactions which take place are

$$E + S \rightleftharpoons ES \rightarrow E + Product.$$

$$K_i \updownarrow +I \qquad (30)$$

$$ESI$$

The overall velocity of this reaction expressed by a Michaelis-Menten type equation is

$$v = \frac{\dfrac{V_{max}[S]}{(1 + [I]/K_i)}}{\dfrac{K_m}{(1 + [I]/K_i)} + [S]} . \qquad (31)$$

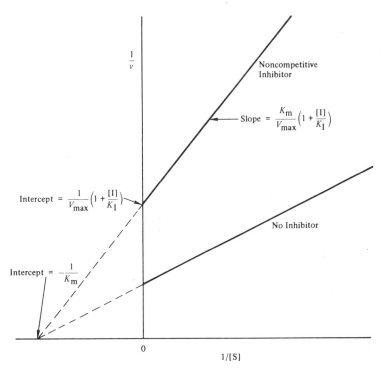

$$\frac{1}{v}$$

Noncompetitive
Inhibitor

Slope $= \dfrac{K_m}{V_{max}}\left(1 + \dfrac{[I]}{K_I}\right)$

Intercept $= \dfrac{1}{V_{max}}\left(1 + \dfrac{[I]}{K_I}\right)$

No Inhibitor

Intercept $= -\dfrac{1}{K_m}$

0

$1/[S]$

Fig. 5-8. Plot of 1/v versus 1/[S] according to the method of Lineweaver and Burk.   Experiments carried out in the absence of and in the presence of a noncompetitive inhibitor.

The Lineweaver-Burk transformation of this equation is

$$\frac{1}{v} = \frac{K_m}{V_{max}}\left(\frac{1}{[S]}\right) + \frac{1}{V_{max}}\left(1 + \frac{[I]}{K_I}\right). \qquad (32)$$

A plot of 1/v versus 1/[S] for this case of uncompetitive inhibition gives a straight line with a slope of K_m/V_{max} (Fig. 5-9).   This line is parallel to the line obtained in the absence of inhibitor.   The vertical intercept is increased to $(1/V_{max})(1 + [I]/K_I)$, and the horizontal intercept is $-(1/K_m)(1 + [I])/K_I$.

Examples of uncompetitive inhibition include the inhibition of the peptidase pepsin by hydrazine, the inhibition of arylsulfatase by

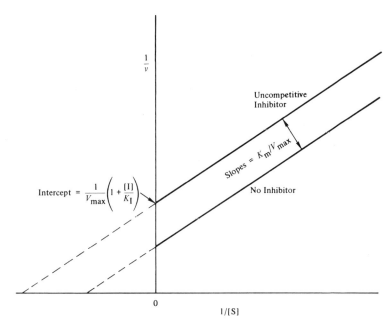

Fig. 5-9. Plot of 1/v versus 1/[S] according to the method of Lineweaver and Burk. Experiments carried out in the absence of and in the presence of an uncompetitive inhibitor.

cyanide or hydrazine, and the inhibition of some dehydrogenases by cyanide, hydroxylamine, or thiols. All of these uncompetitive inhibitors have the common characteristic of being good nucleophiles. Consequently, one might suspect the formation of a covalent bond in the ES—I complex.

The three examples of enzyme inhibition presented above are theoretical models to which experimental data may be compared. The effect of a constant concentration of inhibitor on the slope of the line in the Lineweaver-Burk plots (Figs. 5-7, 5-8, and 5-9) and upon the intercepts may be used to ascertain if inhibition is competitive, uncompetitive, or noncompetitive. The mathematical treatments allow values of K_I to be determined from the appropriate intercepts or slopes, provided the corresponding value in the absence of inhibitor has been determined and the concentration of inhibitor is known.

d. Substrate Inhibition. Occasionally in a simple enzyme-catalyzed reaction, the dependence of velocity on substrate concentration does not yield a hyperbolic curve; velocity may reach a maximum and then drop off as substrate concentration is increased further. The phenomenon is termed *substrate inhibition*. The simplest mechanism which describes this behavior requires the formation of a Michaelis-Menten complex, followed by the complexation of a second molecule of substrate to form an inactive complex:

$$E + S \rightleftharpoons E \cdot S \rightarrow E + Product. \tag{33}$$

$$\begin{array}{c} \updownarrow + S \\ \downarrow K_I \\ E \cdot S \cdot S \end{array}$$

As substrate concentration increases, velocity increases until a maximum value is reached. But at very high substrate concentration, the ES complex is converted to the inactive ESS complex, and velocity decreases.

The cause of substrate inhibition becomes clearer when one considers the structures of the ES and ESS complexes; these are shown schematically in Fig. 5-10a and 5-10b. In the ES complex (Fig. 5-10a), contact between enzyme and substrate is optimum with a single substrate molecule filling the active site; a productive complex is formed which can go on to product. In the ESS complex (Fig. 5-10b), a portion of one substrate molecule may occupy part of the active site and a portion of a second substrate molecule

 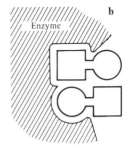

Fig. 5-10. Schematic illustration of the binding of substrate molecules at the active site of an enzyme. **a.** Binding of one molecule in an orientation which allows reaction. **b.** Binding of two molecules in orientations which do not allow reaction and result in substrate inhibition.

may occupy the remainder of the site. This would be expected to result in misorientation of both substrate molecules in a non-productive complex.

The velocity of an enzymatic reaction which shows substrate inhibition, Eq. (33), is given by

$$v = \frac{V_{max}[S]}{K_m' + [S] + K_1[S]^2} \tag{34}$$

where K_1 is, in this case, the equilibrium constant for the formation of ESS from ES and S, and K_m' is a modified Michaelis constant. The reciprocal of Eq. (34) gives an equation analogous to the Lineweaver-Burk equation:

$$\frac{1}{v} = \frac{K_m'}{V_{max}} \left(\frac{1}{[S]} \right) + \frac{1}{V_{max}} + \frac{K_1[S]}{V_{max}}. \tag{35}$$

A plot of $1/v$ versus $1/[S]$ according to this equation is given in Fig. 5-11. At relatively low substrate concentration (i.e., large $1/[S]$), the term $K_1[S]/V_{max}$ is negligibly small, Eq. (35) reduces to the Lineweaver-Burk equation [Eq. (18)], and the plot is linear. However, as substrate concentration becomes large (i.e., at small $1/[S]$), this term becomes significant, and the line curves upward.

Substrate inhibition is shown by a number of single substrate, hydrolytic enzymes. For example, sheep liver carboxylesterase shows this behavior with the substrate ethyl butyrate. However, substrate inhibition is not limited to single substrate reactions. Multiple substrate reactions may show substrate inhibition, and these may proceed by mechanisms more complex than the simple example described here.

pH Dependence of Enzyme-Catalyzed Reactions

The abilities of enzymes to catalyze reactions vary significantly with pH. Enzymes generally have optimal activities at intermediate pH values, and their activities decrease at very high or low pH. Of course, the pH optimum varies from enzyme to enzyme. The pH dependencies of enzymatic reactions can be attributed to either of two general causes. First, pH may affect the stability of an enzyme, causing it to be irreversibly inactivated on either side of its pH optimum. Second, pH may affect the kinetic parameters of the enzymatic reaction. It may affect the stability of the enzyme–

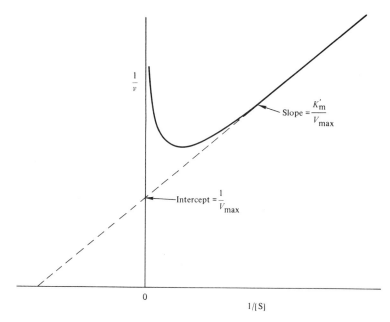

Fig. 5-11. Plot of $1/v$ versus $1/[S]$ according to the method of Lineweaver and Burk for a reaction showing inhibition by substrate.

substrate complex; it may have an effect on the rate of the catalytic step; or it may have an effect on both K_m and V_{max}. This second general cause is of the greatest interest to this discussion, since the kinetically measured constants K_m and V_{max} can be related to the bond-forming and bond-breaking steps of the enzyme-catalyzed reaction.

An immediately striking feature of the pH dependencies of enzymatic reactions is their similarities to those of chemical reactions which are catalyzed by Brønsted acids and bases (Chapter 4). In examining the pH dependencies of enzymatic reactions, as with simple chemical reactions, one attempts to identify the apparent pK_a values of groups which may be involved in catalysis. One would hope to identify those functional groups involved, and their states of ionization. Since enzymatic reactions are more complex than chemical reactions, analyses of the pH dependencies of enzymatic reactions are also more complex. The kinetically apparent

99

pK's may show themselves in the kinetic constants V_{max} or K_m; the actual pK's may be of groups in the enzyme, the substrate, or the enzyme–substrate complex.

Let us consider a model in which the substrate does not ionize, but ionizable groups are present in the enzyme and, consequently, also in the enzyme–substrate complex. The reactive form of the enzyme and the enzyme–substrate complex is the monoionized form of a diacidic species. The overall reaction with its ionization and rate constants is

$$
\begin{array}{ccc}
E & & ES \\
\Big\updownarrow {\scriptstyle +H^+} \Big| K_{e2} & {\scriptstyle +H^+} \Big\updownarrow \Big| K_{es2} & \\
S + EH & \overset{k_1}{\underset{k_2}{\rightleftharpoons}} \ EHS & \overset{k_3}{\rightarrow} EH + Product. \\
\Big\updownarrow {\scriptstyle +H^+} \Big| K_{e1} & {\scriptstyle +H^+} \Big\updownarrow \Big| K_{es1} & \\
EH_2 & EH_2S &
\end{array}
\tag{36}
$$

The velocity of the overall reaction is given by the modified Michaelis-Menten equation

$$
v = \frac{V'_{max}[S]}{K'_m + [S]}
\tag{37}
$$

where V'_{max} is the observed maximum velocity at a constant pH and K'_m is the observed Michaelis-Menten constant at that pH. It may be shown that the observed maximum velocity, V'_{max}, is related to the actual maximum velocity by

$$
V'_{max} = \frac{V_{max}}{1 + [H^+]/K_{es1} + K_{es2}/[H^+]}.
\tag{38}
$$

This equation describes the rate of decomposition of the enzyme–substrate complex as a function of hydrogen ion concentration. It may be easily rearranged to a form closely resembling the equation which describes a bell-shaped dependence of rate constant in a simple chemical reaction (Chapter 4).

As in simple chemical reactions, the evaluation of rate constants and kinetically apparent pK values is done most simply with a plot of $\log V'_{max}$ versus pH (Fig. 5-12a). Since $V'_{max} = k_3[E]_{total}$, Eq. (11), a plot of $\log k_3$ versus pH may also be used. This graph, which follows from Eq. (38), may be broken into linear segments having slopes of $+1$, 0, and -1. As with a simple chemical reaction, the

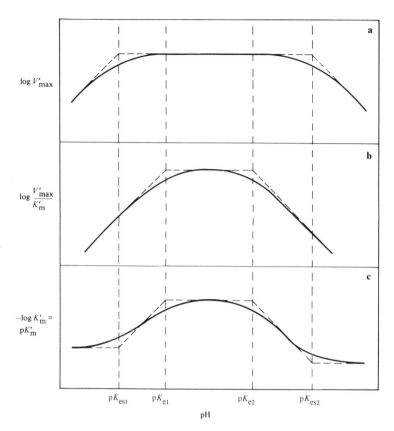

Fig. 5-12. pH dependencies of kinetic constants of enzyme-catalyzed reactions. **a.** Log V'_{max} plotted versus pH; **b.** Log (V'_{max}/K'_m) plotted versus pH; **c.** $-$ Log K'_m plotted versus pH. (Adapted from K. J. Laidler and P. S. Bunting, *The Chemical Kinetics of Enzyme Action,* Clarendon Press, Oxford, 1973, p. 158.)

change of slope from $+1$ to 0 as pH increases suggests the necessity for a Brønsted base in the catalytic step; the change of slope from 0 to -1 suggests the necessity for a Brønsted acid. The pH values at which these linear segments intersect, when extrapolated, correspond to the pK values of the basic group (pK_{es1} in Fig. 5-12**a**) and the acidic group (pK_{es2} in Fig. 5-12**a**) of the enzyme–substrate complex, respectively.

The kinetically apparent pK values obtained by this graphical method are those of groups within the enzyme–substrate complex. Usually, they are shifted significantly from their normal values in aqueous solution. This is due to shielding of the group from the aqueous environment by the substrate and the protein itself, and a resulting inhibition or promotion of dissociation. Clearly, a method for determining the pK's of these groups in the uncomplexed enzymes is desirable.

It was noted earlier in this chapter that at low substrate concentrations, the velocity of an enzymatic reacton is given by $v = (V_{max}/K_m)[S]$. Under this condition velocity, v, is proportional to substrate concentration, and the term V_{max}/K_m is an apparent first-order rate constant for reaction. Since $V_{max} = k_3[E]_{total}$, Eq. (11), any variation of V_{max}/K_m with pH must reflect ionizations of free enzyme rather than those in the enzyme–substrate complex. For the model described by Eq. (36), the variation of V'_{max}/K'_m with hydrogen ion concentration is given by

$$\frac{V'_{max}}{K'_m} = \frac{V_{max}}{K_m} \frac{1}{1 + [H^+]/K_{e1} + K_{e2}/[H^+]}. \tag{39}$$

A plot of $\log(V'_{max}/K'_m)$ versus pH or $\log(k_3/K_m)$ versus pH also yields linear segments (Fig. 5-12**b**). The pH values at which these linear segments intersect when extrapolated correspond to the pK values of the basic group (pK_{e1} in Fig. 5-12**b**) and the acidic group (pK_{e2} in Fig. 5-12**b**) of the free enzyme. The variation of $\log(V'_{max}/K'_m)$ with pH is important, since it allows identification of the pK's of groups which are necessary for catalysis. Often, these do not appear in plots of $\log V'_{max}$ versus pH, as when the substrate combines only with the correctly protonated form of the enzyme.

The value of K'_m may also vary with pH. Its value as a function of hydrogen ion concentration is obtained by dividing Eq. (38) by Eq. (39) to give

$$K'_m = K_m \frac{1 + [H^+]/K_{e1} + K_{e2}/[H^+]}{1 + [H^+]/K_{es1} + K_{es2}/[H^+]}. \tag{40}$$

This equation contains the dissociation constants of basic and acidic groups of the free enzyme (K_{e1} and K_{e2}) and of the enzyme–substrate complex (K_{es1} and K_{es2}). Therefore, a plot of $-\log K'_m$, or pK'_m, versus pH reflects the pK's of all four of these dissociation constants (Fig. 5-12**c**).

To aid in the interpretation of plots of pK'_m versus pH, such as the example given in Fig. 5-12**c**, Dixon formulated the following simple rules.

1. The graph consists of straight-line sections (if the pK values are sufficiently separated) joined by short, curved parts.
2. The straight portions have integral slopes, i.e., zero or one-unit or two-unit slopes, positive or negative.
3. Each bend indicates the pK of an ionizing group in one of the components (either enzyme or substrate), and the straight portions, when extrapolated, intersect at a pH corresponding to the pK.
4. Each pK produces a change of one unit in the slope.
5. Each pK of a group situated in the ES complex produces an upward bend, i.e., an increase of slope with increase of pH; each group situated in either the free enzyme or the free substrate produces a downward bend, i.e., a decrease of slope with increase of pH.
6. The curvature at the bends is such that the graph misses the intersection point of the neighboring straight parts by a vertical distance of 0.3 units (= log 2); if two pK's occur together the distance is equal to log 3.
7. The slope of any straight-line section is numerically equal to the change of charge occurring in that pH range when the complex dissociates into free enzyme and substrate.

Rules 1 through 4 and 6 may be used with equal validity in interpreting plots of log V'_{max} versus pH, as in Fig. 5-12**a**, and plots of log(V'_{max}/K'_m) versus pH, as in Fig. 5-12**b**.

The acidic and basic groups most likely to participate in catalysis and be responsible for the pH dependencies of enzyme-catalyzed reactions are the side-chain functional groups of the amino acids which make up the protein structures of enzymes. The normal pK_a values of these groups are summarized in Table 5-1. By comparison of the kinetically apparent pK's with the known pK_a's of Table 5-1, one would hope to be able to identify the functional groups involved in catalysis. Occasionally, confirmation of the identification of the group or groups involved is sought by determining the enthalpy of ionization, ΔH^0, from the kinetically apparent pK values at a series of temperatures (see Chapter 3 for the method). The ionizable groups found in proteins have characteristic ΔH^0 values which are also summarized in Table 5.1. Furthermore, by determining the ionization states of these groups, one would hope to be able to deduce their mechanistic functions, i.e., general acid, general base or nucleophile. Of course in enzyme-catalyzed reactions, as in chemically catalyzed reactions, the possibility of more than one kinetically equivalent mechanism exists. Therefore, the pH dependence of an enzyme-catalyzed reaction may suggest a catalytic mechanism, but it does not prove the mechanism.

A few examples should illustrate the applicability of the pH dependence of an enzyme-catalyzed reaction to the determination of its mechanism.

a. Ribonuclease. Ribonuclease is synthesized in the pancreas and secreted into the gut where it catalyzes the hydrolysis of RNA during the digestive process. The reaction it carries out is the cleavage of the phosphodiester bond at the 3'-hydroxyl of pyrimidine nucleotides. Ribonuclease also hydrolyzes 2',3'-cyclic phosphodiesters of these residues; therefore, this cyclic structure is considered to be an intermediate in the overall reaction. The pH dependencies of $\log(k_3/K_m')$ for the hydrolyses of several substrates of ribonuclease, including dinucleotides and cyclic phosphodiesters, are shown in Fig. 5-13. The pH dependencies of

Table 5-1. Characteristic, pK_a and ΔH^0 values of acidic and basic groups found in proteins[a,b]

Group		pK_a (25°C)	ΔH^0 (kcal/mole)
α-Carboxyl (terminal)		3.0–3.2	0 ± 1.5
β-Carboxyl (aspartic)		3.0–4.7	0 ± 1.5
γ-Carboxyl (glutamic)	approx.	4.4	0 ± 1.5
Imidazolium (histidine)		5.6–7.0	+6.9–7.5
α-Amino (terminal)		7.6–8.4	+10–13
Sulfhydryl or thiol (cysteine)	approx.	8–9	+6.5–7.0
ε-Amino (lysine)		9.4–10.6	+10–12
Phenolic hydroxyl (tyrosine)		9.8–10.4	+6
Guanidinium (arginine)		11.6–12.6	+12–13

[a] J. T. Edsall in *Proteins, Amino Acids and Peptides* E. J. Cohn and J. T. Edsall, Eds., Reinhold, New York, 1943, p. 445.

[b] Note that the following groups are neither protonated nor ionized to a significant extent, nor have pK_a values in the range from 2 to 13: the peptide bond, the side-chain amides of asparagine and glutamine, the indole of tryptophan, and the alkyl hydroxyl groups of serine and threonine.

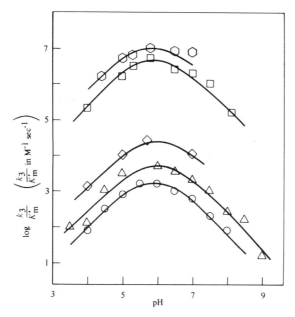

Fig. 5-13. Log(k_3/K'_m) versus pH for the hydrolyses of several substrates by ribonuclease where $V'_{max} = k_3[E]_{total}$. Substrates: 2′,3′-cyclic UMP (○); 2′,3′-cyclic CMP (△); UpU (◇); UpA (□); and CpA (○). (Figure is redrawn from E. J. del Rosario and G. G. Hammes, *Biochemistry* **8**, 1884–1889 (1969)).

these reactions suggest the participation of an apparent general acid and an apparent general base or nucleophile which have pK_{es}'s in the uncomplexed enzymes of 6.4 and 5.4, respectively. Chemical modification studies and X-ray crystallographic analysis indicate that these groups are the imidazole sidechains of histidine-12 and histidine-119, respectively.

A possible catalytic mechanism for the hydrolysis of RNA by ribonuclease is shown in Fig. 5-14. This grossly oversimplified mechanism shows the hydrolysis process taking place in two steps. In the first step the protonated imidazolium cation of histidine-12 functions as a general acid by donating a proton to the oxygen of the alcohol which is to be displaced. Simultaneously, the free base form of the imidazole of histidine-119 acts as a general base by abstracting a proton from the 2′-hydroxyl of the ribose ring;

105

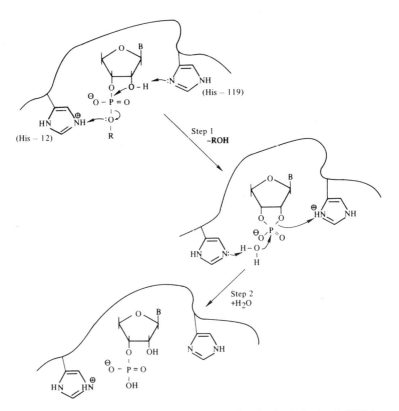

Fig. 5-14. A possible catalytic mechanism for the hydrolysis of RNA by ribonuclease.

the oxygen of the 2'-hydroxyl bonds to the phosphorous atom to displace the alcohol group and form a cyclic phosphodiester intermediate. It should be noted that an identical cyclic phosphodiester is formed as an intermediate in the hydroxide ion-catalyzed hydrolysis of RNA. In the second step of the hydrolysis mechanism a molecule of water assumes the position previously occupied by the alcohol which has been displaced. Histidine-12 now acts as a general base by abstracting a proton from water; the oxygen of the water molecule forms a bond to the phosphorus atom. Simultaneously, histidine-119 functions as a general acid by donating a proton to the 2'-oxygen of the ribose; cleavage of the bond between this oxygen and the phosphorous atom occurs at this time.

b. Lysozyme. A second example which we may consider is the mechanism of action of lysozyme, an enzyme which catalyzes the hydrolysis of the polysaccharide cell walls of gram-positive bacteria. While this enzyme is widely distributed, lysozyme from hens' egg white is the easiest to obtain, and it has been studied in the greatest detail. Lysozyme hydrolyzes glycosidic linkages of alternating polymers of N-acetylglucosamine and N-acetylmuramic acid which are linked $\beta(1 \to 4)$ (Fig. 5-15). However, in kinetic studies of the enzyme, $\beta(1 \to 4)$-linked oligomers of N-acetylglucosamine have been the most useful substrates.

The pH dependence of the log of k_3, which is calculated from $V'_{max} = k_3[E]_{total}$, for the hydrolysis of the hexamer of N-acetylglucosamine (Fig. 5-16a) suggests the nature of those groups within the enzyme–substrate complex which are essential for catalysis. They are an apparent general acid with a pK_{es2} of 6.7 and an apparent general base or nucleophile with a pK_{es1} of 3.8. A plot of $\log(k_3/K'_m)$ versus pH for this same reaction (Fig. 5-16b) indicates that in the free enzyme these groups have pK's of 6.1 and 4.2, respectively. X-ray crystallographic analysis and chemical studies of lysozyme indicate the presence of two amino acid side-chain functional groups at the catalytic site: glutamic acid-35 and aspartic acid-52. The side-chain carboxyl group of glutamic acid-35 is in a hydrophobic environment and probably has a relatively high pK_a; it

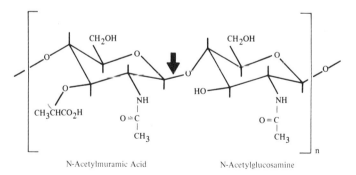

N-Acetylmuramic Acid   N-Acetylglucosamine

Fig. 5-15. The substrate of lysozyme. Arrow indicates the $\beta(1 \to 4)$ glycosidic bond which is hydrolyzed.

107

would function catalytically as the general acid. The side-chain carboxyl group of aspartic acid-52 is in a hydrophilic environment and probably has a relatively low pK_a; it would be in the dissociated form during catalysis. However, there is no evidence to indicate that it functions as either a general base or nucleophile. It should also be noted that in the enzyme–substrate complex, the ring structure of the substrate molecule is distorted from the chair to the half-chair conformation (see Fig. 5-17). It has been proposed that this distortion facilitates catalysis, since the substrate conformation is made to resemble that of the presumed transition state.

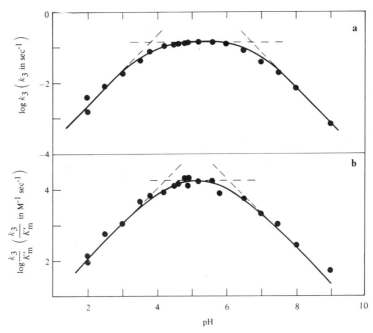

Fig. 5-16. pH dependencies of kinetic parameters for the lysozyme-catalyzed hydrolysis of the $\beta(1 \rightarrow 4)$-linked hexamer of N-acetyl-*D*-glucosamine. **a.** Log k_3 as a function of pH; k_3 was calculated from $V'_{max} = k_3[E]_{total}$. **b.** Log (k_3/K'_m) as a function of pH. (Figures are redrawn from S. K. Banerjee, I. Kregar, V. Turk, and J. A. Rupley, *J. Biol. Chem.* **248**, 4786–4792 (1973).)

A generally accepted mechanism for the catalysis of the hydrolysis of the polysaccharide by these two carboxyl groups is shown in Fig. 5-17. The reaction catalyzed by lysozyme may be

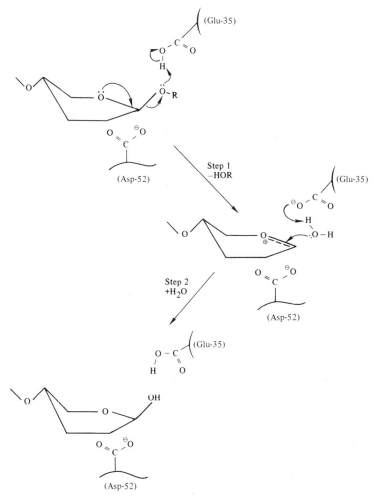

Fig. 5-17. A possible mechanism for the hydrolysis of a glycosidic bond by lysozyme.

viewed as a two-step process. In the first step the undissociated carboxyl of glutamic acid-35 acts as a general acid by donating a proton to the oxygen of the glycosidic bond. Cleavage of the bond between this oxygen and the C-1 carbon atom of the carbohydrate ring can then take place to form an oxonium ion intermediate. As mentioned previously, the dissociated form of the carboxyl of aspartic acid-52 is implicated in the mechanism by kinetic studies; however, there is no evidence to indicate that it acts either as a general base or as a nucleophile. The best guess as to its function is that its negative charge electrostatically stabilizes the positively charged oxonium ion intermediate. In the second step a water molecule assumes the position previously occupied by the alcohol which has been displaced. Glutamic acid-35 now acts as a general base by abstracting a proton from water. The oxygen of the water molecule simultaneously bonds to the C-1 carbon of the hexose, thereby forming the product and regenerating the enzyme.

c. Chymotrypsin. The mechanism of catalysis by chymotrypsin, while relatively simple, is far more complicated than the two preceding examples. Indeed, the details of this reaction are hotly debated. Nevertheless, certain features of the catalytic site and the reaction mechanism may be discussed with a fair degree of certainty. The kinetic studies described earlier in this chapter indicate that the catalytic process involves acylation and deacylation of the enzyme by the carboxylic acid portion of the substrate. The pH dependencies of both acylation and deacylation steps suggest the involvement of a general base or nucleophile with a kinetically apparent pK_{app} of approximately 7. A plot of log k versus pH for the deacylation of acetylchymotrypsin gives the pK_e of this group in the covalent intermediate as 7.3 (Fig. 5-18). A quick examination of Table 5-1 would lead to the conclusion that this group is probably an imidazolium side-chain group of a histidyl residue. Chemical modification studies have shown that histidine-57 is essential for activity, and, in addition that the hydroxyl of serine-195 is the site of acylation. X-ray crystallographic analysis implicates the participation of a third group, the carboxyl of aspartic acid-102, which is dissociated.

The arrangement of the catalytic triad at the active site of chymotrypsin, the position of the substrate, and possible mechanisms of acylation and deacylation are shown in Fig. 5-19. As the substrate is absorbed, the enzyme forces rotation of the peptide bond from its normal conformation in which the carbonyl oxygen and amide

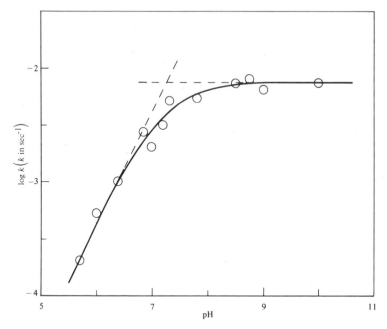

Fig. 5-18. Log of the first-order rate constant of the deacylation of acetyl-chymotrypsin as a function of pH. (Figure is redrawn from F. C. Wedler, F. L. Killian, and M. L. Bender, *Proc. Nat. Acad. Sci. US* **65**, 1120–1126 (1970).)

proton are *trans*. This conformational change prevents resonance stabilization of the peptide bond and makes it more labile. It has been proposed that the arrangement of the carboxylate ion, imidazole group, and hydroxyl group in the catalytic triad exists for the purpose of making the hydroxyl of serine-195 a powerful nucleophile. Aspartate-102 abstracts a proton from histidine-57, which in turn abstracts the proton from the hydroxyl of serine-195. Simultaneously, the oxygen of the hydroxyl, which has been un-usually nucleophilic, attacks the carbonyl carbon of the substrate. The tetrahedral intermediate which is formed loses the amine after it is protonated, possibly by histidine-57. The deacylation reaction essentially follows the reverse course, with water assuming the place of the amine which has been released in the acylation step.

Analyses of the pH dependencies of enzyme-catalyzed reactions

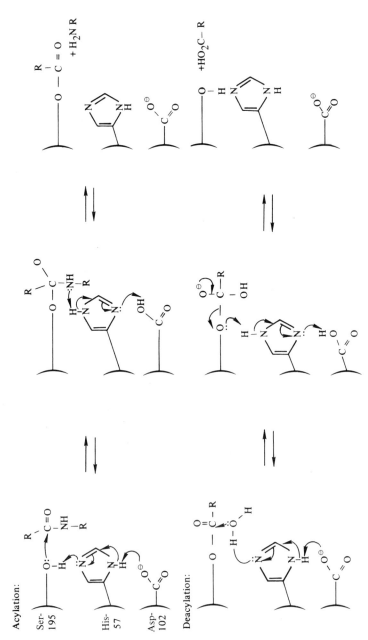

Fig. 5-19. A possible mechanism for the hydrolysis of an amide by chymotrypsin.

for the purpose of deducing catalytic mechanisms have their limitations. Numerous kinetically equivalent rate laws and mechanisms may be proposed to fit the data. Additional information about side-chain functional groups of the amino acids which might be involved should be obtained from other sources, which include chemical modification studies and X-ray crystallographic analysis. Clearly, the pH dependence of an enzyme-catalyzed reaction cannot be used to unambiguously deduce the reaction mechanism, but it can be used to determine what is possible.

Lastly, while participation of Brønsted acids and bases in enzymatic catalysis is extremely important, its significance should be kept in perspective. Other factors have been identified which contribute to catalysis. One of these has already been alluded to in the description of the mechanism of catalysis of lysozyme. An enzyme may strain or distort the conformation of the reactant so that its structure approximates that of the transition state. Other factors which contribute to enzymatic catalysis are implicit in the descriptions of the enzymatic mechanisms presented above. The enzyme must bring catalytic groups and reactant groups into close enough proximity so that bonds may form or be broken. These groups must have the proper spatial orientation to effect maximal catalytic efficiency. Also, the active site of the enzyme provides a microscopic environment which favors reaction. It may exclude water and provide a surrounding of polarity, or local dielectric constant, which accelerates the rate of reaction. All of these factors—acid–base catalysis, distortion, proximity, orientation, and environmental effects—are essential to enzymatic catalysis, even though they may not all be detected easily by kinetic analysis.

REFERENCES

Malcolm Dixon and Edwin C. Webb, *Enzymes,* 2nd edition, Chapter IV, Academic Press, New York, 1964.

John Westley, *Enzymic Catalysis,* Part 1, Harper and Row, New York, 1969.

Keith J. Laidler and Peter S. Bunting, *The Chemical Kinetics of Enzyme Action,* 2nd edition, Chapter 3, Claredon Press, Oxford, 1973.

George P. Hess, "Chymotrypsin-Chemical Properties and Catalysis", in *The Enzymes,* Paul D. Boyer, Ed., 3rd edition, Vol. 3, Chapter 7, Academic Press, New York, 1971.

Frederick M. Richards and Harold W. Wyckoff, "Bovine Pancreatic Ribonuclease," in *The Enzymes,* Paul D. Boyer, Ed., 3rd edition, Vol. 4, Chapter 24, Academic Press, New York, 1971.

Taiji Imoto, L. N. Johnson, A. C. T. North, D. C. Phillips, and J. A. Rupley, "Vertebrate Lysozymes," in *The Enzymes,* Paul D. Boyer, Ed., 3rd edition, Vol. 7, Chapter 21, Academic Press, New York, 1972.

PROBLEMS

1. For the Lineweaver-Burk transformation of the Michaelis-Menten equation, Eq. (18), demonstrate that the intercept of the horizontal axis is $-1/K_m$.

2. For enzyme-catalyzed reactions, the approach of observed initial velocity, v, to a maximum value, V_{max}, at high substrate concentrations is generally interpreted as indicating that an enzyme–substrate complex is formed prior to reaction. This effect also results from the reaction scheme,

$$E \cdot S \overset{K_d}{\rightleftharpoons} E + S \overset{k_2}{\rightarrow} E + \text{Product,}$$

in which $E \cdot S$ is a nonreactive complex with a dissociation constant of K_d; and enzyme, E, and substrate, S, react bimolecularly with a second-order rate constant of k_2 to yield product.

    a. For this alternate reaction scheme, demonstrate that

$$v = \frac{k_2 K_d [E]_{total}[S]}{K_d + [S]}.$$

Use the definitions of $[E]_{total}$, $[S]$, and $[E \cdot S]$ given for the Michaelis-Menten derivation in this chapter. Also assume that $[S]$ is much greater than $[E]_{total}$.

    b. What is the value of $[S]$ when $v = V_{max}/2$?

3. Ribonuclease catalyzes the hydrolysis of o-nitrophenyl oxalate (T. C. Bruice, B. Holmquist, and T. P. Stein, *J. Am. Chem. Soc.* **89**, 4221 (1967)). The formation of o-nitrophenolate as a function of time is shown in the figure at the top of page 115.

    a. What does the shape of this curve suggest about the reaction mechanism?

    b. Ribonuclease is believed to have at its catalytic site two imidazole side-chain functional groups of two histidine residues. At the pH where ribonuclease catalyzes the hydrolysis of o-nitrophenyl oxalate, one of these would be in the conjugate acid

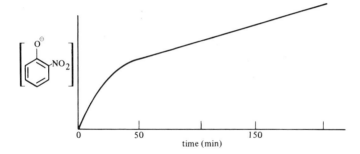

time (min)

form (ImH$^+$) and the other in the free base form (Im:). Suggest a reasonable mechanism for this reaction, showing bonds being formed and broken.

4. Carboxypeptidase A, an enzyme isolated from the pancrease, hydrolyzes amino acids from the carboxy terminus of polypeptides. It removes phenylalanine from the synthetic substrate benzoylgly-cylglycyl-L-phenylalanine (Bz-Gly-Gly-Phe), and this reaction is inhibited by β-phenylpropionate (D. S. Auld and B. L. Vallee, *Biochemistry* **9**, 602 (1970)). Observed rates obtained in the absence and in the presence of 10^{-4}M β-phenylpropionate at several concentrations of Bz-Gly-Gly-Phe are as follows:

Bz-Gly-Gly-Phe (M)	No inhibitor, v (arbitrary units)	$+10^{-4}$M β-phenylpropionate, v (arbitrary units)
2.5×10^{-5}	3.0×10^3	1.55×10^3
5×10^{-5}	4.9×10^3	2.5×10^3
10^{-4}	7.1×10^3	3.7×10^3
2×10^{-4}	9.1×10^3	4.5×10^3

What are the V_{max}, V_{max}/K_m, and K_m values for this reaction? Does the inhibitor act competitively, noncompetitively, or uncompetitively? What is the value of K_I, the dissociation constant of the enzyme–inhibitor complex?

5. Fumarase catalyzes the hydration of fumarate to malate:

$$^{\ominus}O_2C-CH=CH-CO_2^{\ominus} + H_2O \rightarrow HO-CH(CO_2^{\ominus})-CH_2-CO_2^{\ominus}$$

115

This enzymatic reaction is inhibited by succinate. Initial velocities obtained at a constant pH of 6.5 in the absence and in the presence of 5×10^{-2}M succinate at several concentrations of fumarate are in the accompanying table.

[Fumarate] (M)	No inhibitor, v (arbitrary units)	$+5 \times 10^{-2}$M succinate, v (arbitrary units)
5×10^{-5}	0.95	0.57
10^{-4}	1.43	0.95
2×10^{-4}	2.00	1.40
5×10^{-4}	2.50	2.13

a. Determine if the inhibition is competitive, noncompetitive, or uncompetitive. What is the value of K_I, the dissociation constant of the enzyme–inhibitor complex?

b. Given the accompanying tabular data for the hydration of fumarate, what conclusions can be drawn about the functional group(s) which might participate in catalysis? Assume that the substrate is in the di-anionic form over the entire pH range studied (Data adapted from D. A. Brant, L. B. Barnett, and R. A. Alberty, *J. Am. Chem. Soc.* **85**, 2204–2209 (1963)).

pH	V'_{max}/K'_m (min^{-1})
3.0	2.0×10^2
4.0	2.0×10^3
5.0	1.9×10^4
5.7	6.0×10^4
6.4	1.0×10^5
7.1	6.0×10^4
8.0	1.1×10^4
9.0	1.3×10^3

c. The enthalpies of ionization determined from kinetic data for this reaction are: pK_{e1}, $\Delta H^0 = -1.7$ kcal/mole; pK_{e2}, $\Delta H^0 = 7.7$ kcal/mole (Data from D. A. Brant, L. B. Barnett, and R. A. Alberty, *J. Am. Chem. Soc.* **85**, 2204–2209 (1963)). What types of side-chain functional groups of amino acids are most likely to be involved?

d. The data given here are obviously minimal. However, a lack of data has never stopped scientists from speculating on the meaning of the data they have. Can you follow in this tradition by suggesting a catalytic mechanism for fumarase, showing bonds being formed and broken?

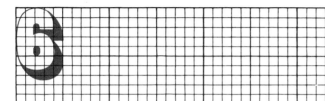

MULTIPLE
SUBSTRATE
REACTIONS

To this point, we have considered only those enzyme-catalyzed reactions which involve either one substrate or one substrate and water as the second substrate. In reality, most enzymes of biological significance catalyze reactions of two or more substrates.

In early studies of enzyme-catalyzed reactions which involved two substrates, it was assumed that the substrates formed a complex with the enzyme in a manner totally independent of each other. Furthermore, it was assumed that the ternary complex thus formed, i.e., the complex of enzyme with both of its substrates, was essential for reaction. The generalized expression which describes this type of reaction mechanism is

$$E + A + B \rightleftharpoons E \cdot A \cdot B \rightleftharpoons E \cdot P \cdot Q \rightleftharpoons E + P + Q \tag{1}$$

where A and B are reactants and P and Q are products. Although some enzymes which catalyze the reaction of two substrates follow this mechanism, not all do. Multiple substrate enzyme mechanisms have been found in which there is an ordered addition of substrates and release of products. Kinetic analyses have been extremely useful in identifying and differentiating among possible mechanisms.

In 1947 Douderoff and his co-workers, in examining the reaction catalyzed by sucrose phosphorylase, demonstrated that the formation of a ternary complex was not necessarily required for a multiple substrate reaction. The overall reaction catalyzed by this enzyme is

$$\text{Glucose-1-P} + \text{Fructose} \rightleftharpoons \text{Sucrose} + P_i. \tag{2}$$

They found that in the absence of the second substrate, fructose, the enzyme could catalyze the incorporation of $[^{32}P]$-phosphate into initially unlabeled glucose-1-phosphate:

$$\text{Glucose-1-P} + {}^{32}P_i \rightleftharpoons \text{Glucose-1-}{}^{32}P + P_i. \tag{3}$$

To explain this result, they proposed the formation of a glucosyl–enzyme intermediate in a two-step reaction:

$$\begin{aligned} \text{Glucose-1-P} + E &\rightleftharpoons \text{E–Glucose} + P_i \\ \text{E–Glucose} + \text{Fructose} &\rightleftharpoons E + \text{Sucrose}. \end{aligned} \tag{4}$$

The incorporation of $[^{32}P]$-phosphate into initially unlabeled glucose-1-phosphate was the result of reversal of the first of these two equilibrium steps.

In 1954 Koshland classified multiple substrate reactions as those which proceed by a *single displacement* mechanism and those that proceed by a *double displacement* mechanism. The former described those reactions which proceed through the ternary complex of enzyme and substrates. The latter described those which proceed through two steps, e.g., the sucrose phosphorylase reaction. The double displacement mechanism could be characterized by the ability of the enzyme to catalyze the exchange of an isotope from one of the products into one substrate in the absence of the second substrate. Furthermore, this mechanism implied an overall retention of steric configuration in going from substrate to product. This retention of configuration was a direct result of an S_N2-type displacement to form a covalent enzyme–substrate intermediate, followed by a second displacement of the enzyme by the second substrate in the step which yielded product. Today it is recognized that isotope exchange and retention of configuration are neither necessary nor sufficient for a double displacement mechanism; however, their presence is suggestive of one.

Since it was not always possible to perform isotope exchange studies or analyses of product conformation in the laboratory, several enzymologists began formulating kinetic analyses of multiple substrate reactions. These mathematical treatments were worked out by King and Altman, Alberty, Dalziel, and Frieden in the late 1950s; and by Hanes and Wong, and Cleland in the early 1960s. While building on the formal, mathematical treatments, Cleland devised more qualitative methods which used the Lineweaver-Burk plot as a tool for distinguishing among the various possible mechanisms.

Terminology

Before considering the various mathematical models of multiple substrate reactions, let us consider the imaginative terminology devised by Cleland to describe them. Two general types of kinetic mechanisms are recognized. First, *Ping Pong* describes a mechanism in which one or more products must be released before all substrates can react. This is the double-displacement mechanism described above. Second, there are *Sequential* mechanisms in which all reactants must combine with enzyme prior to reaction. These are classified further. In *Ordered* mechanisms, reactants combine with the enzyme and dissociate from it in an obligatory order. In *Random* mechanisms the order of combination and release is not obligatory. The number of kinetically important reactants in a given direction are designed by the terms *Uni* (one), *Bi* (two), *Ter* (three), *Quad* (four), etc. The number of these reactants generally excludes water and hydrogen ion. This terminology becomes clearer if one considers a few examples of mechanisms. In these examples, substrates are termed A, B, C, etc; products are termed P, Q, R, etc.; and a modified form of the enzyme (in the double displacement, Ping Pong mechanism) is termed E'.

In the reaction catalyzed by sucrose phosphorylase (described above) one product, phosphate, must be released before the second substrate, fructose, can react, Eq. (4). Therefore, this reaction follows a Ping Pong mechanism. The reaction is bimolecular in reactants and bimolecular in products; hence, it is termed Bi Bi. The overall Ping Pong Bi Bi reaction may be generalized by the equation,

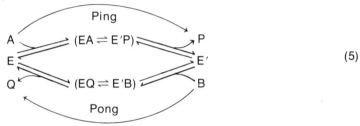

$$(5)$$

Diagrammatic representations of reaction mechanisms can become very complicated; therefore, Cleland also introduced a simplified method for diagraming reaction mechanisms. The pathway of the reaction is indicated as a horizontal line. Generally, above the line, arrows indicate the points of addition of substrates (downward into the line) and release of product (upward from the line). Generally, below the line, letters indicate the enzyme form present at various points along the pathway; and interconversions of species bound to the enzyme are given in parentheses. According to these conventions, the generalized Ping Pong Bi Bi mechanism is represented as

$$(6)$$

Rate constants and equilibrium constants may be associated with the steps of this reaction as with any multiple-step or enzyme-catalyzed reaction.

Many enzymes have been found to catalyze reactions according to the Ping Pong Bi Bi mechanism. They include transaminases and redox reactions of flavoproteins as well as the reaction of sucrose phosphorylase described above.

Enzymes which catalyze reactions which proceed via a Sequential mechanism may be Ordered or Random. A reaction which has an ordered addition of two substrates and ordered release of two products has an Ordered Bi Bi mechanism. This type of reaction follows the generalized reaction mechanism,

$$E \rightleftharpoons EA \rightleftharpoons (EAB \rightleftharpoons EPQ) \rightleftharpoons EQ \rightleftharpoons E. \atop +A \quad +B \qquad\qquad +P \quad +Q \tag{7}$$

According to the conventions of Cleland, the Ordered Bi Bi reaction is

$$\frac{\begin{array}{ccccc} A & B & & P & Q \\ \downarrow & \downarrow & & \uparrow & \uparrow \end{array}}{E \quad EA \quad (EAB \rightleftharpoons EPQ) \quad EQ \quad E}. \tag{8}$$

Many NAD and NADP requiring dehydrogenases catalyze reactions by Ordered Bi Bi mechanisms.

If substrates attach to an enzyme in a manner completely independent of each other, the enzyme mechanism is Random. A Random mechanism which has two substrates and two products is Random Bi Bi. This mechanism is generalized as

$$E \overset{\nearrow EA \searrow}{\underset{\searrow EB \nearrow}{}} (EAB \rightleftharpoons EPQ) \overset{\nearrow EP \searrow}{\underset{\searrow EQ \nearrow}{}} E. \tag{9}$$

The Random Bi Bi reaction mechanism may also be expressed as,

$$\tag{10}$$

This reaction mechanism is used by some dehydrogenases and kinases.

These three multiple substrate reaction mechanisms (Ping Pong Bi Bi, Ordered Bi Bi, and Random Bi Bi) are the simplest, and the kinetic treatments which will differentiate them are considered in detail below. Of course, more complicated mechanisms have been found for other enzymes. A few examples which illustrate the terminology are as follows:

$$\text{Random Uni Bi} \tag{11}$$

$$\text{Ordered Ter Bi} \tag{12}$$

$$\text{Bi Uni Uni Uni Ping Pong} \tag{13}$$

$$\text{Hexa-Uni Ping Pong} \tag{14}$$

In recent years a *modus operandi* has been developed to determine, in a relatively simple manner, the orders of substrate binding and product release in multiple substrate enzyme-catalyzed reactions. Kinetic experiments are used first to determine if the reaction proceeds by a Ping Pong or a Sequential mechanism. After this initial segregation, the order of substrate binding—or lack of it—may be determined by studies which include additional kinetic analyses utilizing inhibitors, isotopes exchange studies, and equilibrium-binding studies. In the following pages these methods

123

of operation are examined in reference to the three simplest multiple substrate reaction mechanisms (Ping Pong Bi Bi, Ordered Bi Bi, and Random Bi Bi). However, it should be realized that these methods of analysis can be—and are—used to determine components of reaction mechanisms more complicated than these three.

Ping Pong Mechanisms*

The initial discrimination between Ping Pong and Sequential mechanisms is based on the features of the Lineweaver-Burk type double reciprocal plots for the enzyme under study. For the Ping Pong Bi Bi mechanism the Lineweaver-Burk type equation is as follows:

$$\frac{1}{v} = \frac{K_m^A}{V[A]} + \frac{K_m^B}{V[B]} + \frac{1}{V}. \tag{15}$$

where v is the observed initial velocity, K_m^A is the Michaelis constant of substrate A, and K_m^B is the Michaelis constant of substrate B; for the sake of simplifying equations, the maximum velocity is generally given as V rather than V_{max} as used in simpler enzymatic reactions. (For a derivation of this equation, the reader is directed to the references listed at the end of this chapter.) Since there are two substrates whose concentrations may be varied, there are two possible double reciprocal plots: $1/v$ versus $1/[A]$ at constant concentrations of B, and $1/v$ versus $1/[B]$ at constant concentrations of A. These plots are given in Figs. 6-1**a** and 6-1**b**, respectively. Plots of $1/v$ versus $1/[A]$ at different constant concentrations of B (Fig. 6-1**a**) yield parallel lines; that is, the slopes of all lines are the same and equal to K_m^A/V. The plots of $1/v$ versus $1/[B]$ at different constant concentrations of A (Fig. 6-1**b**) also yield

---

*The symbols used here and in the following pages to describe the various equilibrium constants are those recommended by the Enzyme Commission. However, the terminology used by Cleland is also in common use. The relationships of these terms are as follows:

Meaning	Enzyme Commission	Cleland
Michaelis constant for A	K_m^A	K_a
Michaelis constant for B	K_m^B	K_b
Dissociation constant for A	K_s^A	K_{ia}
Maximum velocity (V_{max})	V	V_1

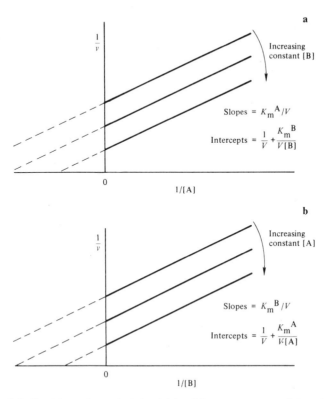

Fig. 6-1. Double reciprocal plots obtained for an enzyme-catalyzed reaction functioning by a Ping Pong mechanism. **a.** Plots of $1/v$ versus $1/[A]$ at several constant concentrations of B. **b.** Plots of $1/v$ versus $1/[B]$ at several constant concentrations of A.

parallel lines with slopes of K_m^B/V. Parallel lines on such double reciprocal plots are diagnostic for a Ping Pong type mechanism. However, great caution should be used in deducing a Ping Pong mechanism from double reciprocal plots which seem to give parallel lines (Fig. 6-1). The sets of lines may converge when extrapolated to a distance and may only seem to be parallel. Use of other methods to confirm the deduction of a Ping Pong mechanism is very desirable. Two straightforward methods are to demonstrate isotopic exchange in the half reaction and to demonstrate

125

that the enzyme also yields parallel double reciprocal plots when assayed in the reverse reaction.

The parallel nature of the lines on these plots is reminiscent of the parallel lines observed on double reciprocal plots obtained from experiments performed in the presence of uncompetitive inhibitors (see Chapter 4). For the case of uncompetitive inhibition, parallel lines resulted from withdrawal of free enzyme, E, from the system in the form of an intermediate, $E \cdot S \cdot I$. The parallel lines of the double reciprocal plot which characterize the Ping Pong mechanism result directly from regeneration of free enzyme, E, from the altered enzyme form, E', by increasing concentrations of the second substrate. Thus, for uncompetitive inhibition and reactions following a Ping Pong mechanism, parallel lines on double reciprocal plots are the result of a modulation of the concentration of free enzyme, E.

The primary plots of $1/v$ versus $1/[A]$ or $1/[B]$ (Figs. 6-1**a** and 1**b**) are useful in indicating the operation of a Ping Pong mechanism. However, they cannot be used directly to evaluate the kinetic and binding constants, since both slopes and intercepts are complex constants containing two unknowns. Secondary plots of the data must be employed. In the plots of $1/v$ versus $1/[A]$ at various constant concentrations of B (Fig. 6-1**a**), the intercepts of the vertical axis are

$$\text{Intercept} = \frac{1}{V} + \frac{K_m^B}{V[B]}. \qquad (16)$$

A plot of the Intercept values versus $1/[B]$ yields a straight line with a slope of K_m^B/V (Fig. 6-2**a**). On this secondary plot the vertical intercept is $1/V$ and the horizontal intercept is $-1/K_m^B$. Similarly, in the plots of $1/v$ versus $1/[B]$ at various constant concentrations of A (Fig. 6-1**b**), the intercepts of the vertical axis are,

$$\text{Intercept} = \frac{1}{V} + \frac{K_m^A}{V[A]}. \qquad (17)$$

A secondary plot of the Intercept values versus $1/[A]$ gives a straight line with a slope of K_m^A/V (Fig. 6-2**b**). The intercept of the vertical axis on this plot is $1/V$, and the intercept of the horizontal axis is $-1/K_m^A$.

In summary, if plots of $1/v$ versus $1/[\text{substrate}]$ give parallel lines at various fixed concentrations of the second substrate, a Ping Pong mechanism is indicated. Secondary plots of the vertical

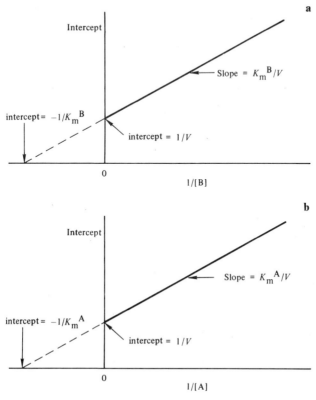

Fig. 6-2. Secondary plots. **a.** Plot of Intercepts of plots of $1/v$ versus $1/[A]$ at fixed concentrations of B (Fig. 6-1**a**) versus $1/[B]$. **b.** Plot of Intercepts of plots of $1/v$ versus $1/[B]$ at fixed concentrations of A (Fig. 6-1**b**) versus $1/[A]$.

intercepts of these plots versus 1/[second substrate] may be used to evaluate the binding and velocity constants which describe the reaction.

Sequential Mechanisms

An enzyme-catalyzed reaction which proceeds via a Sequential mechanism obeys a rate expression somewhat more complicated than that followed by the Ping Pong mechanism. For a Sequential mechanism, the Lineweaver-Burk type equation is as follows:

$$\frac{1}{v} = \frac{1}{V} + \frac{K_m{}^A}{V[A]} + \frac{K_m{}^B}{V[B]} + \frac{K_s{}^A K_m{}^B}{V[A][B]}, \tag{18}$$

where $K_m{}^A$ is the Michaelis constant for substrate A, $K_m{}^B$ is the Michaelis constant for substrate B, and $K_s{}^A$ is the true dissociation constant for substrate A. Since there are two substrates whose concentrations may be varied, there are two possible double reciprocal plots: $1/v$ versus $1/[A]$ at constant concentrations of B, and $1/v$ versus $1/[B]$ at constant concentrations of A. A linear plot of $1/v$ versus $1/[A]$ is obtained when Eq. (18) is rearranged to give

$$\frac{1}{v} = \frac{K_m{}^A}{V[A]} \left(1 + \frac{K_s{}^A K_m{}^B}{K_m{}^A[B]} \right) + \frac{1}{V} \left(1 + \frac{K_m{}^B}{[B]} \right). \tag{19}$$

A linear plot of $1/v$ versus $1/[B]$ is obtained when Eq. (18) is rearranged to give

$$\frac{1}{v} = \frac{K_m{}^B}{V[B]} \left(1 + \frac{K_s{}^A}{[A]} \right) + \frac{1}{V} \left(1 + \frac{K_m{}^A}{[A]} \right). \tag{20}$$

These plots are given in Fig. 6-3**a** and 6-3**b**, respectively.

Several features of these double reciprocal plots should be noted. Both plots of $1/v$ versus $1/[A]$ at constant concentrations of B and plots of $1/v$ versus $1/[B]$ at constant concentrations of A intersect at points to the left of the vertical axis. This characteristic feature is indicative of a Sequential mechanism. One might also note that the points of intersection may be above, below, or on the horizontal axis. If the point of intersection is on the horizontal axis, the binding of the constant substrate has no effect on the K_m of the varied substrate. If the point of intersection is above the horizontal axis, as often happens, the binding of the constant substrate lowers the apparent K_m of the varied substrate. If the point of intersection is below the horizontal axis, the binding of the constant substrate increases the apparent K_m of the varied substrate.

As in the case of the Ping Pong mechanism, the primary plots of $1/v$ versus $1/[A]$ or $1/[B]$ (Figs. 6-3**a** and 6-3**b**) are useful in indicating the operation of a Sequential mechanism. However, since the intercepts and slopes of these lines are complex constants, secondary plots of the data must be used to evaluate kinetic and binding constants. Let us first consider the plot of $1/v$ versus $1/[A]$ at

128

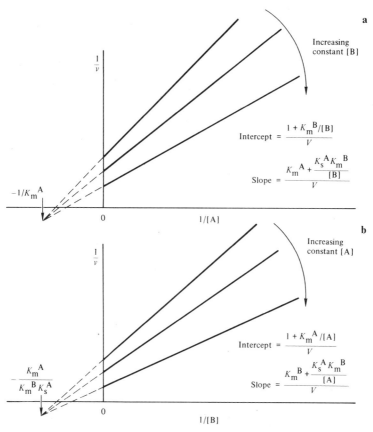

Fig. 6-3. Double reciprocal plots obtained for an enzyme-catalyzed reaction functioning by a Sequential mechanism. **a.** Plots of $1/v$ versus $1/[A]$ at several constant concentrations of B. **b.** Plots of $1/v$ versus $1/[B]$ at several constant concentrations of A.

constant concentration of B (Fig. 6-3**a**). The vertical intercepts are,

$$\text{Intercept} = \frac{1}{V} + \frac{K_m{}^B}{V}\left(\frac{1}{[B]}\right). \tag{21}$$

A secondary plot of the Intercept value versus $1/[B]$ (Fig. 6-4**a**) yields a straight line with a slope of $K_m{}^B/V$. The vertical intercept is

129

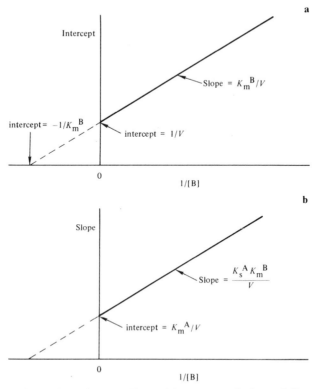

Fig. 6-4. Secondary plots. **a.** Plots of Intercepts of plots of $1/v$ versus $1/[A]$ at fixed concentrations of B (Fig. 6-3**a**) versus $1/[B]$. **b.** Plots of Slopes of plots of $1/v$ versus $1/[A]$ at fixed concentrations of B (Fig. 6-3**a**) versus $1/[B]$.

$1/V$, and the horizontal intercept is $-1/K_m^B$. The slopes of the plots of $1/v$ versus $1/[A]$ (Fig. 6-3**a**) are,

$$\text{Slope} = \frac{K_m^A}{V} + \frac{K_s^A K_m^B}{V} \left(\frac{1}{[B]} \right). \tag{22}$$

A secondary plot of the Slope value versus $1/[B]$ (Fig. 6-4**b**) yields a straight line with a slope of $K_s^A K_m^B / V$ and a vertical intercept of K_m^A / V. Values of V and K_m^B are derived from the vertical and horizontal intercepts of the secondary plot of Intercepts versus $1/[B]$ (Fig. 6-4**a**); K_m^A is calculated from the ratio of the intercepts of the two secondary plots (Figs. 6-4**a** and 6-4**b**); and $K_s{}^A$ is calculated from the ratio of the slopes of these two secondary plots (Figs. 6-4**a** and 6-4**b**). The Intercepts and Slopes of plots of $1/v$ versus $1/[B]$ at constant concentrations of A (Fig. 6-3**b**) may be used in an analogous manner to evaluate these same constants.

Inhibition of Sequential Mechanisms

Once it has been established by the simple kinetic tests outlined above that an enzyme-catalyzed reaction involving two substrates proceeds via a Sequential mechanism, one may ask if the substrates bind to the enzyme in a Random or an Ordered fashion. Inhibitor studies are very useful in distinguishing between these two possibilities. The use of inhibitors to make this distinction is based on the fact that Random and Ordered mechanisms exhibit different patterns of inhibition. The three general types of inhibition found with Sequential reactions are the same as those seen with enzymatic reactions involving a single substrate; these are competitive inhibition, noncompetitive inhibition, and uncompetitive inhibition (Chapter 5).

Three types of compound which are commonly used in inhibition studies are identified as *product inhibitors, alternate product inhibitors,* and *dead-end inhibitors.* A compound that is a normal product of an enzymatic reaction, and inhibits by combining only with the enzyme form it would react with as a substrate in the reverse reaction, is called a product inhibitor. An inhibitor which could have been a normal product produced at a reasonable rate if an alternate substrate had been used, and which combines only with the enzyme form it would combine with as a substrate in a reverse reaction, is called an alternate product inhibitor. The enzyme–inhibitor complexes formed by reaction of product inhibitors or alternate product inhibitors are normal intermediates in the reaction sequence; they can be transformed into other enzyme forms occurring earlier in the reaction sequence. An inhibitor which is never a normal product, but which can combine with enzyme so that the enzyme–inhibitor complex cannot go on to form product is a dead-end inhibitor. It inhibits by lessening the amount of enzyme present which can actually be involved in catalysis. The kinetic effects of product inhibitors and dead-end inhibitors are considered here.

Cleland has formulated a set of rules based on rigorous mathematical analyses for predicting the type of inhibition (i.e., competitive, noncompetitive, or uncompetitive) to be expected by inspection of a proposed mechanism. These rules may be applied to a proposed mechanism unless it contains random sequences which are not in rapid equilibrium. If kinetic constants are to be calculated from the inhibition data, or if nonrapid equilibrium random sequences are involved, one must derive the full rate equation. Cleland's rules are as follows:

Rule 1: The vertical intercept of a double reciprocal plot is affected by an inhibitor which combines with a form of the enzyme other than the one with which the variable substrate combines.

Rule 2: The slope of a double reciprocal plot is affected by an inhibitor which binds to an enzyme form that is the same as —or is connected by a series of reversible steps to—the form of the enzyme with which the variable substrate combines.

The applicability of these rules can be seen most readily when they are considered in the context of the inhibition of single substrate reactions (see Chapter 5). If, in a given reaction scheme where an inhibitor is present, only Rule 1 applies, the inhibition is uncompetitive. If only Rule 2 can be applied to the reaction scheme, competitive inhibition is seen. If both Rules 1 and 2 are applicable, the inhibition is noncompetitive.

When the above rules are applied to enzymatic reactions involving multiple substrates, one additional consideration is taken into account:

Note 1: An inhibitor cannot bind to an enzyme form whose steady-state level is already zero because of saturation by substrate.

We will consider first the inhibitory patterns that product inhibitors would produce in an Ordered Bi Bi reaction. (Inhibitory patterns produced by product inhibitors in a number of the more common multiple substrate reactions are summarized in Table 6-1.)

$$
\begin{array}{cccc}
A & B & P & Q \\
\downarrow & \downarrow & \uparrow & \uparrow \\
\hline
E & EA & (EAB \rightleftharpoons EPQ) & EQ \quad E
\end{array}
\qquad (8)
$$

Table 6-1. Product inhibition patterns[a]

Mechanism	Product inhibitor	Vary A			Vary B			Vary C		
		Unsat	Sat B	Sat C	Unsat	Sat A	Sat C	Unsat	Sat A	Sat B
Ordered Bi Bi	P	NC	UC		NC	NC				
	Q	Comp	Comp		NC	—				
Random Bi Bi	P	Comp	—		Comp	—				
	Q	Comp	—		Comp	—				
Random Bi Bi (nonrapid equilibrium)	P	NC	NC		NC	NC				
	Q	NC	NC		NC	NC				
Ping Pong Bi Bi[b]	P									
	Q									
Ordered Ter Ter	P	NC	UC	UC	NC	NC	UC	NC	NC	NC
	Q	UC	UC	UC	UC	UC	UC	UC	UC	UC
	R	Comp	Comp	Comp	NC	—	NC	NC	—	UC
Bi Uni Uni Ping Pong	P	NC	UC	—	NC	NC	—	Comp	Comp	Comp
	Q	UC	UC	UC	UC	UC	UC	NC	NC	NC
	R	Comp	Comp	Comp	NC	—	NC	UC	—	UC
Hexa-Uni Ping Pong	P	NC	—	NC	Comp	Comp	Comp	UC	UC	—
	Q	UC	UC	—	NC	NC	—	Comp	Comp	Comp
	R	Comp	Comp	Comp	UC	—	UC	NC	—	NC
Bi Bi Uni Uni Ping Pong	P	NC	UC	NC	NC	NC	NC	UC	UC	UC
	Q	UC	UC	—	UC	UC	—	Comp	Comp	Comp
	R	Comp	Comp	Comp	NC	—	NC	NC	—	NC

[a] Adapted from Kent M. Plowman, *Enzyme Kinetics*, McGraw-Hill, New York, 1972, pp. 154–155. Abbreviations: Unsat, unsaturated with; Sat, saturated with; Comp, competitive; NC, noncompetitive; UC, uncompetitive.

[b] In Problem 4 at the end of this chapter, the reader is asked to predict the product inhibition patterns for this reaction; he may enter them into this table after predicting them.

Let us begin with conditions under which the variable substrate is A, B is not at a saturating concentration, and the product inhibitor is Q. Rule 1 does not apply since the variable substrate A and product inhibitor Q combine with the same enzyme form (E in Eq. (8)); therefore, no effect on the intercept is seen. Rule 2 does apply, however, and an effect on slope is seen. Competitive inhibition results (Fig. 6-5**a**). In every other case (Q versus B, P versus A, or P versus B) Rule 1 does apply since inhibitors and substrates bind to different enzyme forms (E rather than EA or EA rather than E); therefore, an effect on intercept is seen. Also, Rule 2 applies since inhibitors bind to forms which are connected by reversible steps to the form of enzyme with which the varied substrate combines; an effect on slope is seen. Since both Rules 1 and 2 may be applied, noncompetitive inhibition is seen in these three cases (Figs. 6-5**b**, 6-5**c**, and 6-5**d**).

A slightly different product inhibition pattern is produced if the substrate held at a constant concentration is also at a saturating level (figure not shown). The difference is seen where the variable substrate is A, B is at a saturating level, and the product inhibitor is P. Rule 1 may be applied since substrate A and inhibitor P combine with different enzyme forms, E and EQ, respectively; therefore, an intercept effect is seen. However, Rule 2 cannot be applied since these two forms are no longer connected by a series of reversible steps; according to Note 1, the product inhibitor P cannot bind to EA (or EQ) because their levels are reduced to zero due to the saturating level of B. The connection is broken by the high concentration of B, and no slope effect is seen. Inhibition in this case is noncompetitive. In the case where B is saturating, A is the varied substrate, and Q is the product inhibitor, the application of Rules 1 and 2 remains the same and noncompetitive inhibition is still predicted. Also when A is saturating, B is varied, and P is the product inhibitor, noncompetitive inhibition is predicted. However, when B is saturating, A is varied, and Q is the product inhibitor, no inhibition is expected. This is due to the fact that according to Note 1, inhibitor Q cannot bind to enzyme because the enzyme is already saturated with A.

The reader should note that the inhibition patterns for the Ordered Bi Bi mechanism may be used to infer the order of binding of substrates to the enzyme. A competitive pattern is seen only when the inhibitor is the last product to be released (i.e., Q), and the varied substrate is the first to be added (i.e., A). Also, an uncompetitive pattern is produced only when the second substrate to bind is held constant (i.e., B), the inhibitor is the first product to be released (i.e., P), and the varied substrate (i.e., A) is the first to be added.

The Random Bi Bi mechanism produces its own characteristic product inhibition patterns which may be predicted by Cleland's

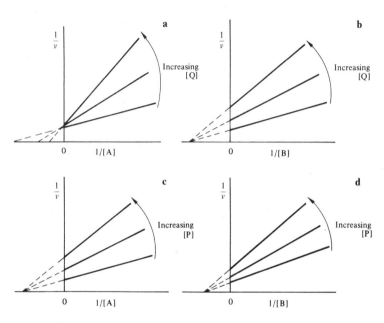

Fig. 6-5. Representative double reciprocal plots showing effects of product inhibitors on an Ordered Sequential mechanism.

a. Pattern given by several concentrations of product inhibitor Q when [A] is varied and [B] is constant.

b. Pattern given by several concentrations of product inhibitor Q when [B] is varied and [A] is constant.

c. Pattern given by several concentrations of product inhibitor P when [A] is varied and [B] is constant.

d. Pattern given by several concentrations of product inhibitor P when [B] is varied and [A] is constant but not saturating. This pattern becomes uncompetitive when [A] is constant and saturating.

In this figure noncompetitive inhibition patterns are shown to intersect on the horizontal axis. This is done for simplicity only. The lines generally intersect either above or below the axis.

135

rules (figures not shown). Application of the rules to all combinations of substrates and inhibitors (Q versus A, P versus A, P versus B, and Q versus B) is identical and leads to deduction of the same inhibitory patterns for these four cases. Since all substrates and products combine with the same enzyme form, E, according to Rule 1 no intercept effects are seen. For the same reason, Rule 2 applies, and slope effects are seen. Therefore, if combination of reactants is random, all product inhibition patterns are predicted to be competitive. These predictions, of course, are based on the assumption that dead-end complexes which would result in the observation of one or more noncompetitive patterns are not formed. At saturating concentrations of either substrate, the level of the enzyme form to which all substrates bind, E, becomes zero. According to Note 1, inhibitor can no longer bind to the enzyme. Therefore, the competitive inhibition seen for the Random Bi Bi mechanism may be overcome by saturating levels of substrate.

Clearly, the use of product inhibitors is a valuable tool in elucidating the nature of a multiple substrate reaction. However, they often may not be used, as for example when the reaction is virtually irreversible. In these cases, dead-end inhibitors, which combine with the enzyme at specific substrate binding sites and are competitive inhibitors relative to one substrate, are useful.

Inhibition patterns produced by dead-end competitive inhibitors may also be predicted by applying Rules 1 and 2. (Inhibitory patterns produced by dead-end competitive inhibitors in a number of the more common multiple substrate reactions are summarized in Table 6-2.) For a two-substrate Sequential reaction in which the binding of substrates is Random, a dead-end inhibitor which is competitive with respect to either substrate generally demonstrates noncompetitive inhibition with respect to the other substrate (figure not shown). In the case of an Ordered mechanism, a dead-end inhibitor which is competitive for the substrate which adds to the enzyme first demonstrates noncompetitive inhibition relative to the second substrate; a dead-end inhibitor which is competitive for the second substrate produces uncompetitive inhibition when the initially bound substrate is varied (figure not shown). Application of Note 1 adds to these deductions that the presence of a saturating concentration of a substrate with which the dead-end inhibitor competes abolishes the inhibition. Inhibition patterns produced by dead-end inhibitors, when applied judiciously, can be used to deduce whether a Sequential mechanism is Random or

Table 6-2. Patterns produced by dead-end competitive inhibitors[a]

Mechanism	Inhibitor competitive for substrate	Vary A	Vary B	Vary C
Ordered Bi Bi	A	Comp	NC	
	B	UC	Comp	
Random Bi Bi	A	Comp	NC	
	B	NC	Comp	
Ping Pong Bi Bi	A	Comp	UC	
	B	UC	Comp	
Ordered Ter Ter	A	Comp	NC	NC
	B	UC	Comp	NC
	C	UC	UC	Comp
Ordered Bi Uni Uni Bi Ping Pong	A	Comp	NC	UC
	B	UC	Comp	UC
	C	UC	UC	Comp
Random Bi Uni Uni Bi Ping Pong	A	Comp	NC	UC
	B	NC	Comp	UC
	C	UC	UC	Comp
Hexa-Uni Ping Pong	A	Comp	UC	UC
	B	UC	Comp	UC
	C	UC	UC	Comp

[a] Adapted from H. J. Fromm, *Initial Rate Enzyme Kinetics,* Springer Verlag, New York, 1975, p. 97–102.

Abbreviations: Unsat, unsaturated with; Sat, saturated with; Comp, competitive; NC, noncompetitive; UC, uncompetitive.

Ordered, and may be used to infer the order of substrate addition and product release in an Ordered mechanism.

Up to this point the inhibition of multiple substrate reactions has been treated qualitatively. The thrust of the preceding paragraphs has been to examine patterns of inhibition as tools for distinguishing between Random and Ordered mechanisms. However, as with single substrate reactions, inhibition of multiple substrate reactions may be treated quantitatively for the purpose of determining the dissociation constants, K_i's, for the enzyme–inhibitor complexes. Although the multiplicity of substrates and steps yields extremely complex rate equations for reactions taking place in the presence of inhibitors, these rate expressions can be treated legitimately in a simplified form.

The simplified Lineweaver-Burk type equation which describes competitive inhibition in a multiple substrate reaction is

$$\frac{1}{v} = \frac{K_m^A}{V'} \left(1 + \frac{[I]}{K_I} \right) \frac{1}{[A]} + \frac{1}{V'}, \tag{23}$$

where V' is the maximum velocity obtainable at the constant concentration of B employed, and K_m^A is the apparent K_m of A at the constant concentration of B used. Other terms have the same meanings as used previously. The similarity of this equation to that which describes competitive inhibition in a single substrate reaction is obvious. From examination of Eq. (23), it can be seen that the intercept of the vertical axis is

$$\text{Intercept} = \frac{1}{V'}, \tag{24}$$

and the slopes of these lines are

$$\text{Slope} = \frac{K_m^A}{V'} \left(1 + \frac{[I]}{K_I} \right) = \frac{K_m^A}{V'} + \frac{K_m^A [I]}{V' K_I}. \tag{25}$$

The Lineweaver-Burk type equation which describes noncompetitive inhibition is

$$\frac{1}{v} = \frac{K_m^A}{V'} \left(1 + \frac{[I]}{K_{I \, slope}} \right) \frac{1}{[A]} + \frac{1}{V'} \left(1 + \frac{[I]}{K_{I \, intercept}} \right). \tag{26}$$

Again, the similarity to the equation describing noncompetitive inhibition in a single substrate reaction is obvious. The terms $K_{I \, slope}$ and $K_{I \, intercept}$ are the values of K_I as determined from the slope and intercept, respectively, of a Lineweaver-Burk type plot. The intercepts of the vertical axis are

$$\text{Intercept} = \frac{1}{V'} \left(1 + \frac{[I]}{K_{I \, intercept}} \right) = \frac{1}{V'} + \frac{[I]}{V' K_{I \, intercept}}, \tag{27}$$

and the slopes of these lines are

$$\text{Slope} = \frac{K_m^A}{V'} = \frac{K_m^A [I]}{V' K_{I \, slope}}. \tag{28}$$

The values of $K_{I \, intercept}$ and $K_{I \, slope}$ are to be determined experimentally. They are not necessarily identical since they are often nonequivalent and more complicated functions than implied by Eq. (26).

The Lineweaver-Burk type equation which describes uncompetitive inhibition in a multiple substrate reaction is

$$\frac{1}{v} = \frac{K_m^A}{V'} \left(\frac{1}{[A]}\right) + \frac{1}{V'} \left(1 + \frac{[I]}{K_i}\right). \tag{29}$$

The vertical intercepts of these lines are

$$\text{Intercept} = \frac{1}{V'} \left(1 + \frac{[I]}{K_i}\right) = \frac{1}{V'} + \frac{[I]}{V'K_i}. \tag{30}$$

The slopes of the lines of these plots are

$$\text{Slope} = \frac{K_m^A}{V'}. \tag{31}$$

The dissociation constants of the enzyme–inhibitor complexes, K_i's, are evaluated easily from secondary plots of the Slopes and Intercepts of the primary double reciprocal plots versus inhibitor concentrations. The equations describing the Slopes for competitive and noncompetitive inhibition (Eq. (25) and (28)) are identical. A plot of these Slopes versus inhibitor concentration is linear (Fig. 6-6), and it also is a simple means of evaluating K_i. Similarly, the

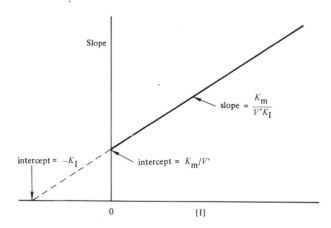

Fig. 6-6. Secondary plot of inhibition data. A plot of the Slopes of Lineweaver-Burk type plots which show competitive and noncompetitive inhibition versus inhibitor concentration.

equations describing the Intercepts of the ordinates for noncompetitive and uncompetitive inhibition (Eq. (27) and (30)) are identical. A plot of the Intercepts versus the inhibitor concentrations is linear (Fig. 6-7), and it serves as a simple means of evaluating K_I. A simple rule-of-thumb for interpreting the secondary plots is that the horizontal intercept of such a plot gives $-K_I$. Alternatively, the ratio of the vertical intercept to the slope of the secondary plot is equal to K_I.

For the sake of simplicity, the above discussion considered only inhibitors which produce linear double reciprocal plots and linear secondary plots of Slopes and Intercepts of the primary plots. However, it is not unusual for the Slope or Intercept replot, or both, to be curved. Where the replot is concave upward, the inhibition is called *parabolic*. Parabolic inhibition results when two or more molecules of inhibitor combine with the enzyme. Where a replot is convex upward, the inhibition is termed *hyperbolic*. It results when inhibition is only partial, that is, the enzyme can still catalyze the reaction, albeit at a reduced rate, when complexed to the inhibitor.

Isotope Exchange Studies

A promising method for distinguishing among the possible multiple substrate reaction mechanisms is the technique pioneered by

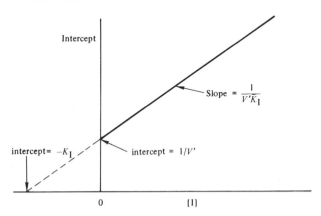

Fig. 6-7. Secondary plot of inhibition data. A plot of the Intercepts of Lineweaver-Burk type plots which show uncompetitive and noncompetitive inhibition versus inhibitor concentration.

Boyer which utilizes the exchange of an isotopically labeled substrate into product. This technique generally monitors the initial velocity of isotope exchange under conditions where substrates and products are in equilibrium. Although quantitative treatments have been devised which allow evaluation of rate constants, this discussion will consider only the use of this technique for distinguishing between the possible multiple substrate reactions.

The use of isotope exchange for identifying a Ping Pong mechanism has already been described. Viewed qualitatively the salient point here is that an isotopic label may exchange between substrate and product in the absence of the second substrate. This feature of the Ping Pong mechanism was discussed amply with the example of sucrose phosphorylase at the beginning of this chapter.

Isotopic exchange studies are also useful for distinguishing between Random and Ordered Sequential mechanisms. Since these studies are generally done at chemical equilibrium, one substrate and one product are usually varied together in constant ratio. The initial velocities of isotopic exchange at varying substrate concentrations are determined, and Lineweaver-Burk type double reciprocal plots are constructed. For the sake of simplicity, we will consider the exchange of isotopically labeled substrate into labeled product (i.e., A* into Q* and B* into P*). However, since an equilibrium process is being studied, the reverse reactions may also be studied with equal validity.

Enzymes which function by the Random Sequential mechanism yield linear double reciprocal plots for the exchange of labeled substrate into product (i.e., A* into Q* and B* into P*) regardless of which substrate, A or B, is varied (Fig. 6-8). This result is intuitively obvious since neither substrate interferes with binding of the other.

A more complex pattern of double reciprocal plots is obtained for enzymes which utilize an Ordered Sequential mechanism. Labeled A* is incorporated into Q* when A is the varied substrate, and labeled B* is incorporated into P* when either A or B are varied at rates which yield linear double reciprocal plots (Figs. 6-9a, 6-9c and 6-9d). However, the double reciprocal plot obtained for the incorporation of A*, the first substrate to bind, into Q* has a pronounced upward curvature at low 1/[B] (i.e., high [B]) (Fig. 6-9b). An upward curvature on a double reciprocal plot is not uncommon, and is generally interpreted as indicating an inhibition of the reaction by the substrate being varied, i.e., substrate inhibi-

141

tion. This effect may be understood qualitatively by considering the reaction mechanism which it describes:

$$\begin{array}{ccccc} A & B & & P & Q \\ \downarrow & \downarrow & & \uparrow & \uparrow \\ \hline E & EA & (EAB \rightleftharpoons EPQ) & EQ & E \end{array} \qquad (8)$$

Increasing the concentration of substrate B (decreasing 1/[B]) shifts the dynamic chemical equilibrium in a way that lowers the amount of free enzyme, E, available to catalyze the exchange of

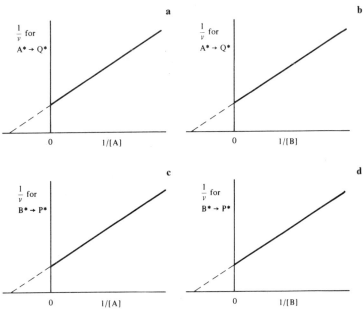

Fig. 6-8. Double reciprocal plots for isotope exchange at equilibrium in a Random Sequential mechanism.

    a. 1/v for exchange of labeled A into Q as a function of 1/[A] at constant [B].

    b. 1/v for exchange of labeled A into Q as a function of 1/[B] at constant [A].

    c. 1/v for exchange of labeled B into P as a function of 1/[A] at constant [B].

    d. 1/v for exchange of labeled B into P as a function of 1/[B] at constant [A].

142

isotopically labeled substrate A* into product Q*. In other words, as [B] increases (1/[B] decreases), the concentration of EAB increases, and the concentrations of EA and free E decrease. Since the amount of E for A* to interact with is lowered, the rate of exchange of A* into Q* must also decrease, and $1/v$ increase.

To summarize, isotope exchange studies may be used to distinguish between Random and Ordered Sequential mechanisms. In addition, this technique may indicate the order of substrate binding in an Ordered mechanism: the substrate whose exchange into product is inhibited by the varied substrate binds first.

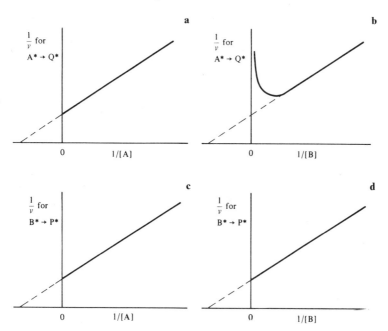

Fig. 6-9. Double reciprocal plots for isotope exchange at equilibrium in an Ordered Sequential mechanism.

a. $1/v$ for exchange of labeled A into Q as a function of 1/[A] at constant [B].

b. $1/v$ for exchange of labeled A into Q as a function of 1/[B] at constant [A].

c. $1/v$ for exchange of labeled B into P as a function of 1/[A] at constant [B].

d. $1/v$ for exchange of labeled B into P as a function of 1/[B] at constant [A].

Substrate Binding Studies

It is generally assumed that before an enzyme-catalyzed reaction takes place, the substrates form complexes with the enzyme which exist in equilibrium with the free species. Equilibrium binding methods are used to examine these initial binding processes. The methodology of these studies is often complex or case specific; therefore, these methods will not be discussed here. However, the results are easily understood; and they are often useful in differentiating between Sequential mechanisms which are Ordered Bi Bi or Random Bi Bi. The Ordered Bi Bi mechanism requires that one substrate bind to the enzyme before the second can bind. Thus, only one substrate should bind in the absence of the second if the mechanism is Ordered Bi Bi, and the substrate which binds is obviously the first to bind. In the Random Bi Bi mechanism, the order of substrate binding is inconsequential. Hence, either substrate should bind in the absence of the other.

This chapter has examined in detail only three possible multiple-substrate reactions: the Ping Pong Bi Bi, the Random Bi Bi, and the Ordered Bi Bi. By comparing experimental data to the mathematical models presented above, one would hope to identify an enzyme mechanism and determine its kinetic and binding constants. Unfortunately, the possibility of kinetic ambiguity, discussed for simple chemical reactions in Chapter 3, exists to an even greater extent for enzyme-catalyzed reactions. Furthermore, this discussion has been somewhat oversimplified for the sake of clarity and brevity. Nevertheless, approaches have been presented which can lead to identification of the enzyme mechanism. While each approach is burdened with its own ambiguities, the use of several of these methods in tandem will progressively eliminate wrong mechanisms. With work and thought, the correct mechanism may be deduced.

REFERENCES

John Westley, *Enzyme Catalysis,* Chapters 7, 8 and 9, Harper and Row, New York, 1969.

Keith J. Laidler and Peter S. Bunting, *The Chemical Kinetics of Enzyme Action,* 2nd edition, Chapter 4, Clarendon Press, Oxford, New York, 1973.

MULTIPLE SUBSTRATE REACTIONS

Kent W. Plowman, *Enzyme Kinetics,* McGraw-Hill, New York, 1972.

Irwin Segel, *Enzyme Kinetics,* Wiley-Interscience, New York, 1975.

W. W. Cleland, "Steady State Kinetics," in *The Enzymes,* Paul D. Boyer, Ed., 3rd edition, Vol. 2, Chapter 1, Academic Press, New York, 1970.

The following papers in the original literature may be of interest:

W. W. Cleland, "The Kinetics of Enzyme-Catalyzed Reactions with Two or More Substrates or Products. I. Nomenclature and Rate Equations," *Biochim. et Biophys. Acta.* **67**, 104–137 (1963).

W. W. Cleland, "The Kinetics of Enzyme-Catalyzed Reactions with Two or More Substrates or Products. II. Inhibition: Nomenclature and Theory," *Biochim. et Biophys. Acta.* **67**, 173–187 (1963).

W. W. Cleland, "The Kinetics of Enzyme-Catalyzed Reactions with Two or More Substrates or Products. III. Prediction of Initial Velocity and Inhibition Patterns by Inspection," *Biochim. et Biophys. Acta.* **67**, 188–196 (1963).

PROBLEMS

1. Chymotrypsin catalyzes the hydrolysis of numerous esters and amides by a mechanism which involves the formation of a covalent, acyl-enzyme intermediate. How would this mechanism be described in the terminology of Cleland? Give a diagrammatic representation of this mechanism.

2. The amino acyl-tRNA synthetases are a group of twenty enzymes which catalyze the esterification of the twenty amino acids to their corresponding tRNA's prior to protein synthesis. These enzyme-catalyzed reactions have three substrates: amino acid, ATP, and tRNA; and three products: amino acyl-tRNA, AMP, and pyrophosphate.

   a. Give diagrammatic representations of all of the possible reaction mechanisms and describe them in the terminology devised by Cleland. (Note: When the author began to work out this problem, he found that like Dr. Frankenstein he had created a monster. He gave up after devising 47 relatively simple diagrammatic representations. This problem serves to illustrate how in science an apparently simple problem can become less simple when one considers all of the possibilities.)

   b. Many investigators believe that the reaction catalyzed by these enzymes proceeds in two steps: (1) amino acid and ATP react to produce pyrophosphate, which is released from the enzyme, and amino acyl-AMP, which remains bound to the enzyme; and (2) reaction of tRNA with enzyme-bound amino acyl-AMP to yield amino acyl-tRNA plus AMP, both of which are released. Which of the possible diagrammatic representations and descriptions from the first part of this problem apply to this mechanism?

3. Most of the twenty amino acyl-tRNA synthetases yield similar plots of $1/v$ versus $1/[substrate]$. Consider the following two cases both at a constant concentration of tRNA: (1) ATP is the varied substrate, and amino acid is the substrate that is fixed at various concentrations; and (2) amino acid is the varied substrate, and ATP is the substrate that is fixed at various concentrations. These cases yield the accompanying double reciprocal plots.

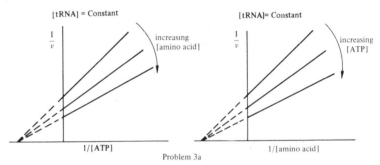

Problem 3a

a. Does the binding of ATP and amino acid occur in a Ping Pong fashion, or is a ternary complex of ATP, amino acid, and enzyme formed?

When the concentrations of ATP or amino acid are held constant, and the concentrations of the other substrates are varied, the double reciprocal plots in the accompanying figure are obtained.

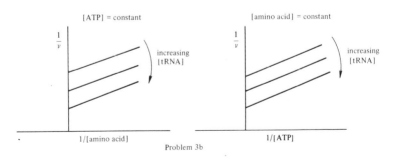

Problem 3b

146

b. What additional conclusions may be drawn about the mechanism of these enzymatic reactions?

4. By applying Cleland's Rules 1 and 2, predict the product inhibition patterns for the Ping Pong Bi Bi mechanism. Assume that neither substrate, when held constant, is at a saturating concentration. Then reapply Rules 1 and 2 and invoke Note 1 to determine the inhibition patterns for this mechanism when each substrate held constant is at a saturating concentration.

5. α-Acetohydroxy acid isomeroreductase, an enzyme common to the biosynthetic pathway for valine, leucine, and isoleucine, catalyzes the conversion of an acetohydroxy acid to an α,β-dihydroxy acid:

$$CH_3-\overset{\overset{O}{\|}}{C}-\overset{\overset{OH}{|}}{\underset{\underset{CH_3}{|}}{C}}-CO_2H + NADPH + H^+ \rightleftharpoons CH_3-\overset{\overset{OH}{|}}{C}-\overset{\overset{OH}{|}}{\underset{\underset{H}{|}}{C}}-CO_2H + NADP^+$$

(α-acetolactate)   (α,β-dihydroxyisovalerate)
                                              (DHIV)

This enzyme yields the double reciprocal plots in the accompanying figure. (Data adapted from E. M. Shematek, S. M. Arfin, and W. F. Diven, *Arch. Biochem. Biophys.* **158**, 132–138 (1973).)

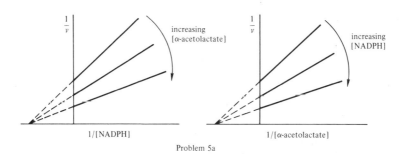

Problem 5a

a. Is the mechanism of this reaction Ping Pong or Sequential?
The reaction catalyzed by this enzyme is inhibited by its products, α,β-dihydroxyisovalerate (DHIV) and NADP. Double reciprocal plots showing the effects of these inhibitors are in the accompanying figure.

147

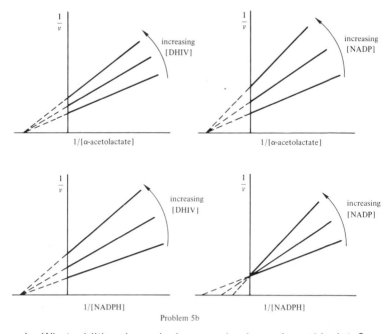

Problem 5b

b. What additional conclusions can be drawn from this data?

6. Inhibitor studies are very useful in determination of the order of substrate binding and Sequential mechanisms. K_i values for the dissociation of inhibitor from the enzyme–inhibitor complexes may be determined from the secondary plots of Lineweaver-Burk plot Slopes and Intercepts versus inhibitor concentrations (Figs. 6-6 and 6-7). Demonstrate that for these secondary plots the horizontal intercepts are equal to $-K_i$.

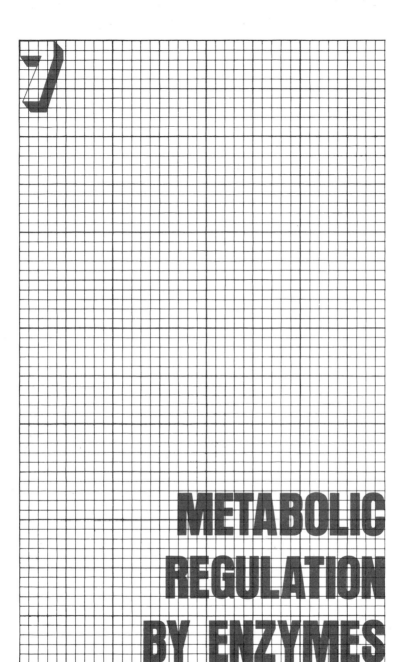

7

METABOLIC REGULATION BY ENZYMES

Since enzymes catalyze reactions within living organisms, it is sometimes desirable for an organism to regulate the rate of product formation by modifying the activity of an enzyme. These regulatory phenomena manifest themselves in two closely related forms. First, the velocity of product formation as a function of increasing substrate concentration may give an S-shaped curve rather than the rectangular hyperbola described by the Michaelis-Menten model (Chapter 5). Second, the enzyme may be inhibited or activated by a compound which is structurally unrelated to either substrate or product. There are many examples of enzyme-catalyzed reactions which are inhibited by the ultimate products of their metabolic pathways. Generally, the enzyme involved catalyzes the first step of the pathway, and it is said to be subject to *end-point inhibition* or *feedback inhibition.* The inhibitors or activators of enzymatic reactions are also termed *modulators* or *effectors. Ligand* is a general term used to describe a substrate, activator, or inhibitor which is assumed to bring about a regulatory effect by binding to the enzyme.

The S-Shaped Dependence of *v* on [S]

The first observation which led to our present understanding of regulatory phenomena was not of an enzyme-catalyzed reaction at all, but of the oxygen binding properties of hemoglobin, the oxygen carrier of blood. While hemoglobin does not catalyze the conversion of oxygen to a product, it does bind oxygen in a manner analogous to the formation of a noncovalent enzyme–substrate complex. A plot of the degree of oxygenation of hemoglobin versus the partial pressure of oxygen is analogous to a plot of velocity versus substrate concentration for an enzymatic reaction. This oxygen binding curve of hemoglobin is S-shaped (Fig. 7-1). However, the oxygen binding curve for myoglobin, the oxygen-carrying protein of muscle, is a simple rectangular hyperbola. The anomalous oxygen binding property of hemoglobin eluded explanation when Bohr first observed it in 1903; however, its physiological significance was obvious. Hemoglobin binds oxygen in the lungs where the partial pressure is high. It flows to the tissues where the partial pressure is low, and it releases oxygen. In the tissues, myoglobin absorbs oxygen and stores it. Myoglobin then releases oxygen as it is needed for biological oxidative reactions.

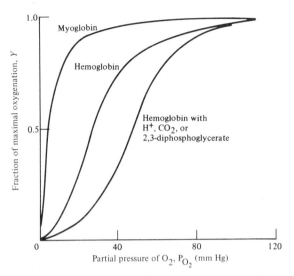

Fig. 7-1. Oxygen binding curves of hemoglobin, myoglobin, and hemoglobin in the presence of three negative heterotropic effectors.

In viewing the S-shaped oxygen binding curve of hemoglobin (Fig. 7-1), one can see that the binding of a small amount of oxygen aids in the binding of additional quantities of oxygen. The phenomenon of ligand binding influencing the binding of additional molecules of the same substance to a protein or enzyme is a *homotropic* effect. If, as in the example of oxygen binding to hemoglobin, initial ligand binding assists further binding or accelerates the observed reacton rate, a *positive homotropic* effect is seen. Dozens of enzymes exhibit positive homotropic effects; and until recently, all homotropic effects were thought to be positively cooperative. We now recognize that the initial binding of a ligand may also inhibit binding of additional molecules of the same ligand; this is a *negative homotropic* effect or *negative cooperativity.*

The binding of a substrate or the velocity of an enzyme-catalyzed reaction may also be modified by a substance that is structurally unrelated to the substrate. Such an effect is *heterotropic.* It may be either positive or negative; that is, the effector may either accelerate or inhibit the reaction. For example, several

152

substances have heterotropic effects on the binding of oxygen by hemoglobin. Hydrogen ion (low pH), carbon dioxide, and 2,3-diphosphoglycerate all lessen the ability of hemoglobin to bind oxygen (Fig. 7-1), and these are negative heterotropic effects. These effects for hydrogen ion and carbon dioxide make sense physiologically, since these substances are produced by active tissues which require oxygen. The negative heterotropic effect of hydrogen ion on the oxygen binding ability of hemoglobin is commonly known as the Bohr effect.

A well-known example of a negative heterotropic effect on an enzyme-catalyzed reaction is the inhibition of the first enzyme of the pathway for isoleucine biosynthesis, threonine deaminase, by the product, isoleucine. A plot of the relative enzyme activity as a function of isoleucine concentration is shown in Fig. 7-2. The curve is S-shaped, indicating that isoleucine not only influences the overall rate of reaction, but also influences its own interaction with threonine deaminase. Low concentrations of the inhibitor have little effect, but inhibition by isoleucine becomes significant above a threshold value. Isoleucine has a negative heterotropic

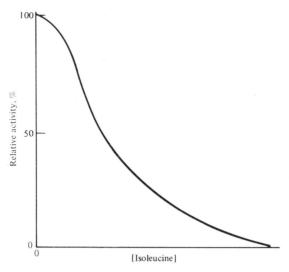

Fig. 7-2. The effect of isoleucine concentration on the relative activity of threonine deaminase.

effect on the velocity of the reaction catalyzed by threonine de-aminase; it has a positive homotropic effect on its own interaction with this enzyme.

While the causes of the S-shaped dependence of the velocities of enzymatic reactions on concentration of substrate or inhibitor are only now becoming understood, their physiological significance is immediately apparent. The organism can finely tune the over-all rate of metabolism according to the concentrations of sub-strates and end products of reaction pathways. For the example of threonine deaminase, low levels of isoleucine have little effect on velocity (Fig. 7-2). However, a supply of isoleucine adequate for the needs of the organism results in inhibition of the first enzyme of the pathway of its synthesis. The net result is operational efficiency: the organism synthesizes product at a high rate when its level is low, but at a low rate when it is not needed.

Activators and Inhibitors

As we have seen in preceding chapters, an enzyme-catalyzed reac-tion may be described by two kinetic constants, the Michaelis-Menten constant, K_m, which is commonly considered to approxi-mate the dissociation constant of the enzyme–substrate complex, and the maximum velocity of the reaction, V_{max}. Regulatory sub-stances, that is, activators and inhibitors, may affect either of these constants. Enzymatic reactions which are subject to regulation are classified as *K systems* or *V systems,* depending on which con-stant, measured kinetically, is altered by the regulator. These terms are generally used in reference to enzymes showing positive homotropic interactions, and examples given in the following dis-cussion describe such cases. However, these terms may also be used to describe enzymes which follow simple Michaelis-Menten kinetics.

The operation of *K*-type and *V*-type regulators is illustrated in plots of reaction velocity as a function of substrate concentration (Fig. 7-3). For the sake of this discussion, *K* is defined as that con-centration of substrate at which half of the maximum reaction velocity is obtained. The presence of a *K*-type inhibitor decreases the affinity of enzyme for substrate, increases the apparent value of *K,* and shifts the curve to the right. Generally, the inhibitor has no effect on the maximum attainable velocity, V_{max} (Fig. 7-3**a**). A *K*-

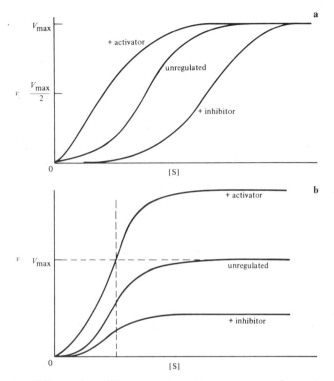

Fig. 7-3. Effects of modifiers on regulatory enzymes. **a.** Dependence of reaction velocity on substrate concentration in the absence of and in the presence of a K-type activator and inhibitor. **b.** Dependence of reaction velocity on substrate concentration in the absence of and in the presence of a V-type activator and inhibitor.

type activator increases the affinity of enzyme for substrate, decreases the apparent value of K, shifts the curve to the left, and might make the curve appear more hyperbolic in shape. It generally has no effect on V_{max} (Fig. 7-3**a**). At most substrate concentrations, the observed velocity, v, of an enzymatic reaction is less in the presence of a K-type inhibitor than in its absence; the observed velocity is greater in the presence of a K type activator than in its absence (Fig. 7-3**a**). The presence of a V-type inhibitor decreases the maximum attainable velocity, V_{max}; a V-type activator increases the value of V_{max} (Fig. 7-3**b**). V-type inhibitors and activators may

155

function without having any effect on the value of K. However, it is also possible for a modifier to act simultaneously as a V-type and a K-type effector.

In summary, an enzymatic activity may be regulated by a homotropic effector (substrate) or by a heterotropic effector (inhibitor or activator). The effect may be positive (by an activator) or negative (by an inhibitor). The effect may be seen as a result of modification of the substrate binding (K system) or as a result of modification of the reaction velocity (V system).

Equilibrium Models

Much of the effort which has been expended to explain regulatory phenomena in enzyme-catalyzed reactions has concentrated on the S-shaped dependence of reaction rate (or ligand binding as in the case of hemoglobin, Fig. 7-1) on substrate concentration, that is, on positive homotropic interactions. Several models which have been proposed to explain this regulatory effect are summarized on the following pages.

a. The Hill Model. In 1909 Hill proposed a simple model which describes the S-shaped binding curve of hemoglobin (Fig. 7-1). More recently, Atkinson and others have adapted the Hill model to describe enzymatic reactions as well. Two basic assumptions are made for the Hill model. First, the free enzyme, E, exists in equilibrium with enzyme which is complexed with n molecules of substrate, S. Second, the concentrations of all complexes containing fewer than n molecules of substrate are negligibly small. Based on these assumptions, the overall enzyme-catalyzed reaction is expressed as,

$$\mathrm{E} + n\mathrm{S} \underset{k_2}{\overset{k_1}{\rightleftharpoons}} \mathrm{ES}_n \overset{k_3}{\rightarrow} \mathrm{E} + n \text{ Product.} \tag{1}$$

The overall velocity of this reaction is given by,

$$v = k_3[\mathrm{ES}_n]. \tag{2}$$

An expression describing the velocity, v, as a function of substrate concentration is obtained in a manner similar to the Haldane derivation of the Michaelis-Menten equation (Chapter 4). The terms used in this derivation are:

$[E]_{total}$ = total concentration of enzyme
$[S]$ = total concentration of substrate
$[ES_n]$ = concentration of enzyme–substrate complex, ES_n
$[E]_{total} - [ES_n]$ = concentration of free enzyme.

As in the Haldane derivation of the Michaelis-Menten equation, the steady-state assumption is made. The rate of formation of ES_n is given by,

$$\frac{d[ES_n]}{dt} = k_1([E]_{total} - [ES_n])[S]^n, \tag{3}$$

and the rate of decomposition of ES_n is given by,

$$-\frac{d[ES_n]}{dt} = k_2[ES_n] + k_3[ES_n]. \tag{4}$$

According to the steady-state assumption, the rate of formation of ES_n is equal to the rate of its decomposition:

$$k_1([E]_{total} - [ES_n])[S]^n = k_2[ES_n] + k_3[ES_n]. \tag{5}$$

Equation (5) is then rearranged to isolate the rate constants:

$$\frac{([E]_{total} - [ES_n])[S]^n}{[ES_n]} = \frac{k_2 + k_3}{k_1} = K'. \tag{6}$$

The factor $(k_2 + k_3)/k_1$ may be expressed as the equilibrium constant K' which is analogous to the Michaelis-Menten constant, K_m. If the rate of decomposition of ES_n back to enzyme and substrate is faster than product formation, k_2 is greater than k_3; and

$$K' = \frac{k_2 + k_3}{k_1} \approx \frac{k_2}{k_1}. \tag{7}$$

Thus, if k_3 is appropriately small or nonexistent, as in the case of hemoglobin, K' is the dissociation constant of the ES_n complex.

Equation (6) also may be rearranged to obtain the concentration of ES_n in terms of substrate concentration:

$$[ES_n] = \frac{[E]_{total}[S]^n}{K' + [S]^n}. \tag{8}$$

This value of $[ES_n]$, when substituted into Eq. (2), gives the observed velocity as a function of substrate concentration:

$$v = \frac{k_3[E]_{total}[S]^n}{K' + [S]^n}. \tag{9}$$

157

Also, since the maximum possible velocity $V_{max} = k_3[E]_{total}$,

$$v = \frac{V_{max}[S]^n}{K' + [S]^n},$$

(10)

or

$$\frac{v}{V_{max}} = \frac{[S]^n}{K' + [S]^n}.$$

(11)

Equations (10) and (11) describe the S-shaped curve characteristic of an enzyme-catalyzed reaction which is influenced by positive homotropic interactions in the binding of substrate. Since v/V_{max} is a measure of saturation of enzymatic activity, it is analogous to the fraction of maximal oxygenation, Y, observed for hemoglobin. In like manner, substrate concentration for an enzymatic reaction is analogous to the partial pressure of oxygen, P_{O_2}. Thus the terms Y and P_{O_2} may be substituted into Eq. (11) to describe the oxygen binding curve of hemoglobin (Fig. 7-1):

$$Y = \frac{(P_{O_2})^n}{K' + (P_{O_2})^n}.$$

(12)

Evaluation of the constants K' and n from an S-shaped plot of v/V_{max} versus [S] is not a simple task. Therefore, a method of plotting this data in linear fashion is desirable. The commonly used linear transformation of Eqs. (11) and (12), when graphed, is the Hill plot. Equation (11), when multiplied by $V_{max}(K' + [S]^n)$, gives,

$$vK' + v[S]^n = V_{max}[S]^n,$$

(13)

or,

$$vK' = [S]^n(V_{max} - v).$$

(14)

Dividing Eq. (14) by $(V_{max} - v)K'$ yields,

$$\frac{v}{V_{max} - v} = \frac{[S]^n}{K'}.$$

(15)

The logarithm of Eq. (15) is the commonly used linear transformation of the Hill equation:

$$\log \frac{v}{V_{max} - v} = n \log [S] - \log K'.$$

(16)

A plot of $\log[v/(V_{max} - v)]$ versus $\log [S]$ is a straight line (Fig. 7-4).

158

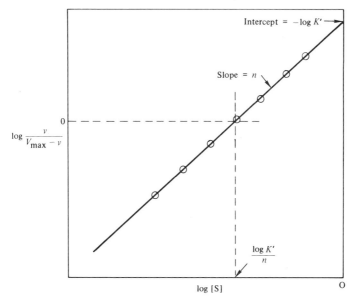

Fig. 7-4. The Hill plot. At $\log[v/(V_{max} - v)] = 0$, $\log K' = n \log[S]$.

The number of molecules of substrate binding to the enzyme in the reactive complex, n, is the slope of this line. The value of K' may be evaluated from the intercept of the vertical axis. However, in practice this intercept can be obtained only by a long extrapolation from the experimental data. This extrapolation is accompanied by a substantial uncertainty; therefore, an alternate method for evaluating K' is used. At half the maximum attainable velocity, $v = V_{max}/2$, and

$$\log \frac{v}{V_{max} - v} = \log \frac{V_{max}/2}{V_{max} - V_{max}/2}. \tag{17}$$

Since

$$\log \frac{v}{V_{max} - v} = \log 1 = 0, \tag{18}$$

at $v = V_{max}/2$,

$$0 = n \log[S] - \log K', \tag{19}$$

159

or

$$n \log[S] = \log K'. \tag{20}$$

Thus, $\log K'$ may also be calculated as $n \log[S]$ where $\log [v/(V_{max} - v)] = 0$ on the Hill plot (Fig. 7-4).

The Hill model resulted in an equation which could be used to describe positive homotropic interactions in enzyme-catalyzed reactions and in the oxygen binding curve of hemoglobin (Fig. 7-1); however, it has several serious, theoretical shortcomings. First, according to the Hill model, the coefficient, n, is equal to the number of ligand or substrate molecules which bind to the enzyme. Generally, this is not the case, and n has a value less than the actual stoichiometry. For example, for the binding of oxygen to hemoglobin, $n = 2.6$ to 2.8, but hemoglobin actually binds four molecules of oxygen. Second, the coefficient n, which is the slope of the Hill plot (Fig. 7-4), is not constant at all substrate concentrations; it asymptotically approaches a value of one at very high and very low substrate concentrations. Third, the model predicts unprecedented and unfeasible kinetics of reaction. For example, the binding of oxygen to hemoglobin is described as requiring a simultaneous interaction of four molecules of oxygen with one molecule of hemoglobin. This pentamolecular reaction would occur in a kinetically fifth-order reaction, an obviously impossible situation.

While the Hill model has theoretical shortcomings, it also has practical advantages which give it a place in contemporary enzymology. The Hill model is a simple, quantitative, empirical relationship which may be used to describe many enzymatic reactions in which there is a nonhyperbolic dependence of velocity on substrate concentration.

In another attempt to describe the oxygen binding curve of hemoglobin (Fig. 7-1), Adair, in 1925, proposed that four molecules of oxygen bind to hemoglobin in four successive steps, each with its own association constant. While this treatment did fit the data accurately, it gave no theoretical explanation for the changing affinity constants. In 1935, Pauling proposed a model which attempted to relate the change in these constants to the geometry of the protein by assuming a single affinity constant and an interaction term which depended on the geometry of the four subunits of hemoglobin. This model foreshadowed the equilibrium models which are in vogue today.

b. The Monod, Wyman, Changeux (MWC or Concerted) Model.

A major impetus to the understanding of enzyme regulatory phenomena came during the 1960s when Perutz and his co-workers reported their X-ray crystallographic findings on the three-dimensional structures of the deoxygenated and oxygenated hemoglobin molecules (Fig. 7-5). The hemoglobin molecule is

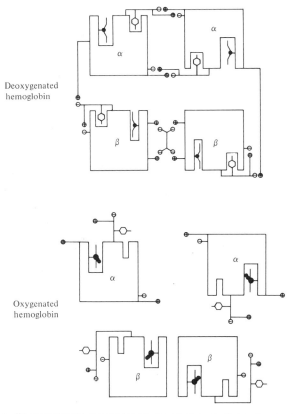

Deoxygenated
hemoglobin

Oxygenated
hemoglobin

Fig. 7-5. Diagrammatic representations of the conformations of the deoxygenated ("tight") and oxygenated ("relaxed") hemoglobin molecule. Deoxygenated hemoglobin has numerous salt bridges; one molecule of 2,3-diphosphoglycerate is held between the β subunits; the four iron atoms are out of the plane of the porphyrin rings; and four tyrosine residues are in hydrophobic pockets. In oxygenated hemoglobin, salt bridges have been broken; 2,3-diphosphoglycerate has been released; the iron atoms are in the plane of the porphyrins; and four tyrosine residues have been removed from the pockets. (M. F. Perutz, *Nature* **228**, 726–739 (1970)).

composed of four polypeptide chains, two designated as α chains and two as β chains. The α and β chains have slightly different amino acid sequences. Each polypeptide is associated with an iron-containing heme group which is the site of oxygen binding. The oxygen binding sites lie far apart from one another, and the deoxygenated and oxygenated forms have detectably different conformations.

In 1965, Monod, Wyman, and Changeux proposed a model which attempts to fuse kinetic and ligand binding data with the newer structural discoveries. They proposed that *indirect* interactions between distinct, specific binding sites are responsible for the performance of regulatory function, and they coined the term *allosteric* to describe these indirect interactions and their effects. Although the model which they proposed primarily offers an interpretation of cooperative homotropic effects, it can be extended to account for heterotropic interactions also.

The model is described by the following statements:

1. Allosteric proteins are oligomers consisting of a number of identical subunits known as protomers.
2. There is one and only one site on each protomer capable of complexing ligand.
3. All of the subunits of a given oligomeric protein are in the same conformation; if one subunit changes its conformation, the others must do so also. All subunits change their conformations in a *concerted* manner.
4. If each subunit can exist in two conformations, there are two forms of the enzyme, E_t and E_r. In the E_t form, all of the subunits are said to be in a "tight" conformation; in the E_r form, all of the subunits are in a "relaxed" conformation.
5. Ligands have a much greater affinity for the E_r conformation than the E_t form. This statement is expressed quantitatively as,

$$K_t = \frac{[E_t][S]}{[E_t S]} \gg K_r = \frac{[E_r][S]}{[E_r S]}. \qquad (21)$$

A schematic representation of this model for a four-subunit protein showing the "tight" E_t form, the "relaxed" E_r form, the equilibrium between these forms, and the equilibria of substrate binding is given in Fig. 7-6. The ratio of protein in the two states in the absence of any ligand is the allosteric equilibrium constant,

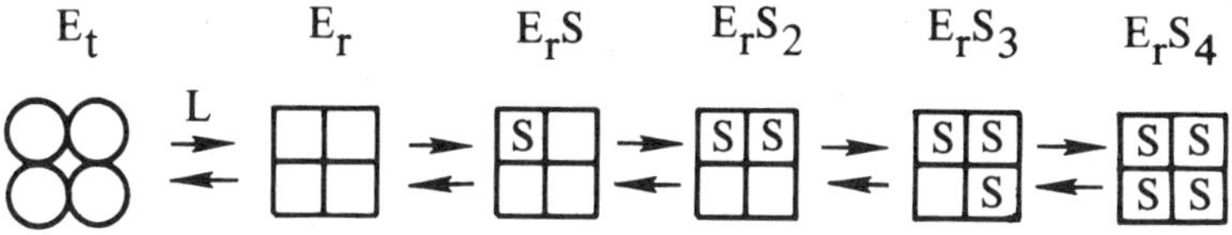

$$E_t \qquad E_r \qquad E_rS \qquad E_rS_2 \qquad E_rS_3 \qquad E_rS_4$$

Fig. 7-6. Schematic representation of the Monod, Wyman, and Changeux concerted model for a four-subunit protein. Tight conformation, $E_t = \bigcirc$; relaxed conformation, $E_r = \square$.

$$L = \frac{[E_r]}{[E_t]}. \tag{22}$$

The *saturation function, Y*, defines the degree of ligand binding or the reaction velocity. The MWC model culminated in an equation which defines the saturation function, *Y*, in terms of the concentration of ligand, S, and the equilibrium constants associated with the system:

$$Y = \frac{Lc\alpha(1 + c\alpha)^{n-1} + \alpha(1 + \alpha)^{n-1}}{L(1 + c\alpha)^n + (1 + \alpha)^n}, \tag{23}$$

where L is the allosteric equilibrium constant, Eq. (22), and the terms α and c are given by,

$$\alpha = \frac{[S]}{K_r} \quad \text{and} \quad c = \frac{K_r}{K_t}. \tag{24}$$

What is most important to know about the MWC equation, Eq. (23) is not the details of its derivation; anyone interested in these may find them in the original paper (see references at the end of this chapter). What is most important is that this model and Eq. (23) describe with reasonable accuracy the positive homotropic interactions of the binding of oxygen by hemoglobin and the velocities of many enzymatic reactions. In doing so the model went beyond describing in mathematical form the experimentally observed regulatory phenomena. It presented a mechanism based on the structures and conformational changes of the proteins themselves. Its greatest contribution was not its correctness in every detail, but its role as a platform upon which more sophisticated models could be based.

c. The Koshland, Nemethy, Filmer (KNF or Sequential) Model. The MWC model broke ground in describing enzymatic regulatory phenomena by introducing the concept that conformational

163

changes of a protein are intimately involved in its function. A somewhat more sophisticated alternate model was proposed by Koshland, Nemethy, and Filmer in 1966. While the MWC-concerted model allowed only two conformations of the protein, the KNF model allowed the existence of intermediate conformational forms. It postulated that a conformational change might be induced in an individual subunit or subunits as a consequence of binding of ligand, and that the conformational changes would take place in a *sequential* manner. Schematic representations of various possible modes of binding the ligand, S, to a tetrameric protein and the consequent sequential conformational changes are illustrated in Fig. 7-7. All of these possibilities are described by the equation,

$$E_{t4} \underset{\substack{-S \\ K_1}}{\overset{+S}{\rightleftharpoons}} E_{t3}E_{r1}S \underset{\substack{-S \\ K_2}}{\overset{+S}{\rightleftharpoons}} E_{t2}E_{r2}S_2 \underset{\substack{-S \\ K_3}}{\overset{+S}{\rightleftharpoons}} E_tE_{r3}S_3 \underset{\substack{-S \\ K_4}}{\overset{+S}{\rightleftharpoons}} E_{r4}S_4. \tag{25}$$

Mathematical treatments describing this model were pioneered by Adair in 1925 and Pauling in 1935. Although these treatments assumed a static protein structure, Koshland, Nemethy, and Filmer adapted them to their model of a protein undergoing conformational changes upon ligand binding. The KNF-sequential model postulates an initial equivalence of binding sites, each with an

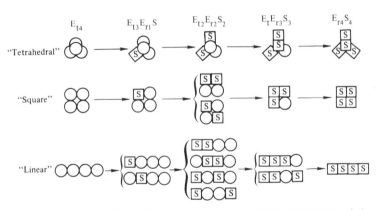

Fig. 7-7. Schematic illustrations of the various modes of binding of the ligand, S, to a tetrameric protein in the Koshland, Nemethy, Filmer sequential model. Three possible geometrical arrangements of the subunits are shown. Tight conformation, $E_t = \bigcirc$; relaxed conformation, $E_r = \square$.

intrinsic dissociation constant for the ligand of K. Binding of the first ligand is described by the dissociation constant $K_1 = K$, binding of the second ligand is described by $K_2 = \alpha K$, binding of the third ligand is described by $K_3 = \beta K$, and binding of the fourth ligand is described by $K_4 = \gamma K$ where,

$$\alpha = 1 - \frac{(1 - c)^2/2c}{1 + (1/2Lc) = (Lc/2)}, \tag{26}$$

$$\beta = 1 - \frac{(1 - c)^2/2c}{1 + (1/2Lc^2) + (Lc^2/2)}, \tag{27}$$

$$\gamma = 1 - \frac{(1 - c)^2/2c}{1 + (1/2Lc^3) + (Lc^3/2)}. \tag{28}$$

The relative velocity of the reaction at any given substrate concentration can then be shown (by anyone with sufficient time, interest, and patience) to be:

$$\frac{v}{V} = \frac{[S]/K[1 + 3\alpha[S]/K + 3\alpha^2\beta([S]/K)^2 + \alpha^3\beta^2\gamma([S]/K)^3]}{1 + 4[S]/K + 6\alpha([S]/K)^2 + 4\alpha^2\beta([S]/K)^3 + \alpha^3\beta^2\gamma([S]/K)^4}. \tag{29}$$

In the context of this discussion there is little point in dwelling on the details of this equation other than to note its complexity. With the vast number of constants included in it, it should not be surprising that it can very precisely describe almost any curve of a positive homotropic interaction. Koshland, Nemethy, and Filmer realized this fact. They noted that while their model could describe almost any set of experimental data, the agreement of the model with the data did not prove the validity of the model. As noted in the preceding chapters, kinetic analysis is subject to ambiguity. Additional information is generally required as final proof of a mechanism.

Lastly, it should be noted that in fitting Eq. (29) to an S-shaped curve which describes a positive homotropic interaction, the tacit assumption is made that $K_1 > K_2 > K_3 > K_4$. This assumption describes a positive cooperative effect such as that seen in Fig. 7-1. However, this assumption need not necessarily be the case. If $K_1 < K_2 < K_3 < K_4$, the observed curve would have a shape which more closely resembled a hyperbola than the S-shape. This would indicate a *negative cooperativity* in ligand binding; binding of one ligand would inhibit binding of additional ligand molecules. Substrate binding curves of intermediate shapes could be generated by using appropriate values for the four dissociation constants. Thus, the KNF-concerted model allows, in theory, a greater variety of shapes of substrate binding curves than either the Hill or MWC-concerted models.

d. The Frieden (Association-Dissociation) Model. The Hill, MWC, and KNF models for positive homotropic interactions are based on the abilities of ligands to bind to oligomeric proteins. However, many enzymes undergo association–dissociation reactions of their subunits, and the kinetic properties of these enzymes (activity or ability to bind ligand) might depend upon the particular state of aggregation. Frieden has proposed a model for the S-shaped dependence of velocity on substrate concentration based on this possibility.

Frieden's model is illustrated most simply with an example. Rabbit muscle phosphorylase a degrades glycogen, a polymeric form of glucose. The enzyme exists as a 380,000 molecular weight species which dissociates reversibly to a 190,000 molecular weight form. The 380,000 molecular weight form binds the substrate glycogen poorly, and can be viewed as being the "tight" form. The dissociated form of the enzyme has a greater affinity for substrate, and is viewed as the "relaxed" form. The association–dissociation model of Frieden is described in a fashion analogous to the MWC-concerted model; the only difference is that the two conformational forms which exist in equilibrium are oligomer and monomer.

e. The Bernhard (Asymmetric Dimer) Model. In recent years it has been realized that many oligomeric proteins and enzymes may be frozen into conformations which are composed of asymmetric dimers, that is, pairs of polypeptides of identical covalent structure which have different tertiary structures. Schematic representations of the simple asymmetric dimer and the analogous tetramer, using the convention of squares and circles, are shown in Fig. 7-8a. Evidence which supports the existence of such structures includes *half of the sites reactivity,* a phenomenon in which only half of the polypeptide chains of an oligomeric protein function catalytically or are modified chemically, presumably due to differences in their tertiary structures. Furthermore, X-ray crystallographic evidence now indicates that asymmetric dimers may be isolated in stable form. Excellent examples of these are the dimeric crystals of insulin and the dimeric holoenzyme forms of liver alcohol dehydrogenase, hexokinase, and malate dehydrogenase.

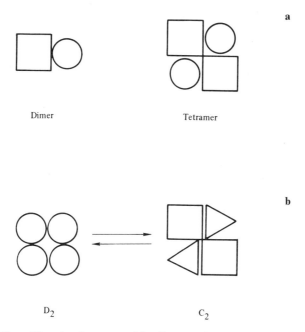

Fig. 7-8. a. The simple asymmetric dimer and an asymmetric tetramer. **b.** Equilibrium between a tetramer of identical subunits (D_2) and a form containing a pair of asymmetric dimers (C_2).

The existence of a stable asymmetric dimer in an enzyme leads directly to the prediction of negative cooperativity or negative homotropic interaction in its kinetic properties. It is reasonable to assume that one monomeric unit may be more catalytically active than the other in a way analogous to that in which one diastereoisomer may react at a rate different from its diastereomer. When considered on the basis of ability to bind substrate, the enzyme subunit with the greater ability will show its greatest catalytic activity at lower substrate concentrations; that conformational form with a lesser ability to bind substrate will exhibit its catalytic activity at higher substrate concentrations. The enzyme will appear to act as an equimolar mixture of enzymes that perform the same function, yet have different K_m values. The net effect is an apparent negative homotropic interaction in a *rigid* asymmetric dimer; apparent positive cooperativity will never be seen.

It is also possible that asymmetric dimers are not held *rigidly* in their conformations. An equilibrium may exist between a form having subunits of identical conformations (D_2 in Fig. 7-8b) and a form containing asymmetric dimers (C_2 in Fig. 7-8b). If, over any substrate concentration range, the C_2 state prodominates, negative cooperativity in ligand binding will be exhibited. If D_2 is the stable conformation of the enzyme, but substrate binds exclusively to the two types of C_2 sites, positive cooperativity will be demonstrated. This should be obvious, since the situation described is analogous to the MWC-concerted model.

It is probably incorrect to view the asymmetric dimer as a static structure. A plausible dynamic model may be conceived in which negative cooperativity is the result of coordinated conformational changes in the two subunits. One subunit would be in an "open" conformation in which it could absorb substrates and release products, but could not catalyze reaction. The other subunit would be in a "closed" conformation in which reaction would occur, but where substrates could not readily be absorbed or product released. The subunits would oscillate back and forth between conformations complexing with substrates, effecting reaction, and releasing products. At low substrate concentrations, only half of the active sites would appear to be functional. Only at relatively high substrate concentrations would binding of substrate and reaction at the remaining sites become significant.

Thus, the concept of the asymmetric dimer is the basis for a plausible explanation of half of the sites reactivity in chemical modification reactions of enzymes, and both positive and negative cooperativity in enzyme kinetics.

Kinetic Models

The models described above have attempted to explain positive homotropic interactions on the variability of affinity of protein for ligand as a function of ligand concentration. While this assumption is valid for the binding of oxygen by hemoglobin (Fig. 7-1), it is not necessarily true for enzymatic reactions. Models have also been proposed which take into account possible kinetic effects.

Rabin has proposed that positive homotropic interactions are possible if the enzyme can exist in two forms of different activities. The overall reaction is:

$$
\begin{array}{l}
\rule{3cm}{0.4pt}\text{E}' + \text{Product.} \\
\text{E} + \text{S} \rightleftharpoons \text{ES} \rightleftharpoons \text{E}'\text{S} \\
\rule{3cm}{0.4pt}\text{E}' + \text{S}
\end{array} \qquad (30)
$$

The intermediate ES cannot give product directly, but E'S can. The model assumes that the interconversion of E' and E is slow relative to all other reactions in the scheme. Thus, when an enzyme molecule E is converted to E' as the result of having been bound to substrate, it "remembers" its active conformation and may easily convert other molecules of substrate S to product. At low concentrations of S, most enzyme is in the form of E or ES, and a slow rate of reaction results. As the concentration of S is increased, the equilibrium is shifted in the direction of the active form of the enzyme, E' or E'S. The net result is the production of an S-shaped dependence of reaction velocity on substrate concentration. Unlike the Hill, MWC, KNF, Frieden, or Bernhard models, the Rabin model does not require multiple catalytic sites or subunits. However, implicit in it, as in the MWC and KNF models, is the necessity for a conformational change of the enzyme to give an active form.

Ferdinand has shown that an S-shaped dependence of reaction velocity upon substrate concentration may be detected in a bisubstrate reaction which proceeds via the following mechanism:

$$
\begin{array}{c}
\quad\; {}^{k_1}\!\!\nearrow \text{EA} \searrow^{k_3} \\
\text{E} \underset{{}_{k_2}\searrow}{\overset{{}^{k_{-1}}}{\rightleftarrows}} \quad \overset{k_{-3}}{\underset{k_4}{\rightleftarrows}} \text{EAB} \rightarrow \text{E} + \text{P} + \text{Q.} \\
{}_{k_{-2}}\;\; \text{EB} \; {}^{k_{-4}}
\end{array} \qquad (31)
$$

It is assumed that formation of the enzyme–substrate complex EAB is slower by the pathway $E \rightarrow EB \rightarrow EAB$ than by the pathway $E \rightarrow EA \rightarrow EAB$; i.e., $k_1 k_3 > k_2 k_4$. While the mechanism described by Eq. (31) is formally Random Bi Bi, it is not completely so, since the pathway via $k_1 k_3$ is favored. If the concentration of either substrate is constant and saturating (i.e., approaches infinity), a plot of velocity versus the concentration of the other substrate will describe a hyperbola. However, under certain assay conditions an enzyme-catalyzed reaction following the mechanism of Eq. (31) can generate an S-shaped dependence of velocity on substrate concentration. To illustrate this, let us consider an experiment in which the concentration of B is fixed at a level well below saturation and the concentration of A is varied. At low concentration of A, most of A will react with EB via the slower pathway described by k_2k_4. But, as the concentration of A becomes greater, more free enzyme, E, will react with A, thereby favoring the dissociation of EB to E plus B. Thus, the slower pathway described by k_2k_4 becomes less important, and the faster pathway described by k_1k_3 predominates. The kinetic result appears to be a positive homotropic interaction. It should be noted that the Ferdinand model, like the Rabin model, does not require multiple catalytic sites or subunits.

¡Experimental Differentiation Among Models

The equilibrium and kinetic models presented above, which account for metabolic regulation at the enzyme level, all may have validity in describing particular systems. The problem facing the experimenter is to determine which of these possible models is functioning for any given system. While analysis of kinetic data is often complex and there is the possibility of ambiguity of interpretation, there are methods which may be used to limit the number of possibilities. Several of these methods are discussed here. For the sake of simplicity, only their applicability in distinguishing among the equilibrium models is considered; these diagnostic tests are summarized in Table 7-1.

a. Plots of *v* versus [S] or *Y* versus [S]. The simplest manner of presenting data is to plot reaction velocity, *v,* versus the substrate concentration, or to plot the extent of saturation by ligand, *Y,* versus ligand concentration. As seen in Chapter 5, a rectangular hyperbola indicates the functioning of the Michaelis-Menten model. For an enzyme with multiple binding sites, these sites do not interact. If such a plot yields an S-shaped or sigmoid curve, either the Hill, MWC-concerted, KNF-sequential, Frieden, or Bernhard models might adequately describe the data. A plot of *v* versus substrate concentration, or a plot of *Y* versus ligand concentration which does not describe a single smooth sigmoid curve but shows other multiple curved segments or "bumps," requires a complex mechanism. Deviation from an S-shaped curve (or "bumps") is generated as the result of initial binding of ligand, inhibiting binding of additional molecules of ligand. As noted

Table 7-1. Summary of diagnostic tests for regulatory models

Test	Observation	Conclusion	Equilibrium models allowed
Plot of v vs. [S] or Y vs. [S]	Hyperbolic curve	Noninteracting sites	—
	S-shaped (sigmoid) curve	Positive cooperativity	Hill, MWC, KNF, Frieden, Bernhard
	"Bumps"	Possibility of positive and negative cooperativity	KNF, Bernhard
Plot of $1/v$ vs. $1/[S]$ or $1/Y$ vs. $1/[S]$	Straight line	Noninteracting sites, V and K_m determined	—
	Concave upward	Positive cooperativity	Hill, MWC, KNF, Frieden, Bernhard
	Concave downward	Negative cooperativity	KNF, Bernhard
Hill plot	$n = 1$	Noninteracting sites	—
	$n =$ number of subunits	Positive cooperativity	Hill
	$1 < n <$ number of subunits	Positive cooperativity	MWC, KNF, Frieden, Bernhard
	$n < 1$	Negative cooperativity	KNF, Bernhard
Scatchard plot	Straight line	Noninteracting sites	—
	Concave downward	Positive cooperativity	Hill, MWC, KNF, Frieden, Bernhard
	Concave upward	Negative cooperativity	KNF, Bernhard

above, negative cooperativity such as this is consistent only with the KNF-sequential or Bernhard-asymmetric dimer models.

b. Double Reciprocal Plots. The second treatment of kinetic data which is generally performed is a Lineweaver-Burk type plot of $1/v$ versus $1/[S]$ or $1/Y$ versus $1/[S]$. A straight line on such a plot indicates that the Michaelis-Menten mechanism is in operation (Chapter 5); and, for an enzyme with multiple sites of ligand binding, these sites do not interact. The S-shaped dependence of v or Y on substrate concentration, when plotted in double reciprocal form, yields a curve which is concave upward and has a linear asymptote. This result is consistent with the Hill, MWC-concerted, KNF-sequential, Frieden, and Bernhard models which describe a positive cooperativity in ligand binding. If, however, a line is obtained which is concave downward, negative cooperativity of ligand binding is implied. (Such a curve is also generated by the presence of enzyme molecules with different K_m values.) Negative cooperativity is consistent only with the KNF-sequential and Bernhard asymmetric dimer models.

c. Hill Plots. It was noted earlier in this chapter that one of the shortcomings of the Hill model, Eq. (1), is that the coefficient n, which is the slope of the Hill plot (Fig. 7-4), is not constant at all substrate concentrations; it asymptotically approaches a value of one at very high and very low substrate concentrations. These segments of the curve can now be understood as the contribution to the reaction rate by the extreme conformations acting independently. To illustrate, if one considers the MWC-concerted model, at low substrate concentrations only the "tight" form is present to bind substrate. The Hill coefficient is one because there is a simple association of one substrate molecule with one binding site of the "tight" form. Also, at high substrate concentration only the "relaxed" state has free sites, and only a simple association is seen. However, at intermediate substrate concentrations, a transition is seen between the "tight" and "relaxed" forms. The slope of the segment of the Hill plot which corresponds to these concentrations is n, the observed Hill coefficient, and it is viewed as an index of the degree of cooperativity between subunits.

While the Hill model, Eq. (1), in most cases does not accurately describe the cause of positive homotropic interactions, the Hill plot (Fig. 7-4) is useful as a diagnostic aid in distinguishing among the possible equilibrium models. In those few cases where n, which is the slope of the Hill plot (Fig. 7-4), is equal to the number of subunits, the Hill model *appears* to be obeyed. That is, the enzyme goes from E to ES_n without building up the intermediates ES_{n-1}, ES_{n-2}, etc. This is an extreme case of the MWC-concerted model in which only E_t and E_rS_n are permitted. An enzyme which appears to obey the Hill model is deoxycytidine monophosphate deaminase which has four subunits and $n = 4$.

If n, as obtained from the Hill plot, is equal to one, a simple equilibrium is indicated in which there is no interaction between binding sites or subunits. However, if n is greater than one but less than the number of subunits, some interaction of ligand binding sites and subunits is indicated. In this case the empirical Hill coefficient, n, represents a combination of the number of sites and the degree of interaction between them. The MWC-concerted, the KNF-sequential, the Frieden, and Bernhard models are consistent with such a value of n. Finally, a value for n of less than one is obtained when binding of an initial molecule or molecules of ligand inhibits binding of additional ligands; negative cooperativity is indicated. Thus, a value for n of less than one is consistent only with the KNF-sequential and Bernhard asymmetric dimer models.

d. The Scatchard Plot. To this point we have considered analyses of kinetic data which might be used to distinguish among the possible equilibrium models proposed to explain homotropic interactions. Binding data may also be used for this purpose if the concentrations of ligand and enzyme–ligand complex can be measured at equilibrium. The use of equilibrium binding data was first employed by Scatchard to determine the number of ligand binding sites, n, and the intrinsic *association* constants when there is *no interaction* of binding sites.

The Scatchard treatment begins with the assumption of a simple association equilibrium between binding sites, B, and ligand or substrate, S, to yield complex, BS:

$$B + S \rightleftharpoons BS. \tag{32}$$

The association equilibrium which defines the reaction, assuming no interaction of sites or cooperativity of ligand binding, is

$$K = \frac{[BS]}{[B][S]}. \tag{33}$$

The terms [S] and [BS] are measurable quantities. The concentration of uncomplexed binding sites, B, is given by, $[B] = [B]_{total} - [BS]$. If there are n binding sites per enzyme molecule, then

173

$$[B]_{total} = n[E]_{total}, \tag{34}$$

and

$$[B] = n[E]_{total} - [BS]. \tag{35}$$

Substituting this concentration of B into Eq. (33) yields

$$K = \frac{[BS]}{(n[E]_{total} - [BS])[S]}, \tag{36}$$

or

$$K(n[E]_{total} - [BS]) = \frac{[BS]}{[S]}, \tag{37}$$

and

$$\frac{[BS]}{[S]} = Kn[E]_{total} - K[BS]. \tag{38}$$

This is a linear algebraic expression, and when [BS]/[S] is graphed as a function of [BS], it yields the Scatchard plot (Fig. 7-9). The intercept of the vertical axis is $Kn[E]_{total}$, the slope of the line is $-K$. The intercept of the horizontal axis, where [BS]/[S] = 0, is

$$0 = Kn[E]_{total} - K[BS], \tag{39}$$

or

$$K[BS] = Kn[E]_{total}, \tag{40}$$

and

$$[BS] = n[E]_{total}. \tag{41}$$

Thus, if the concentrations of BS and S can be determined, the Scatchard plot may be used to determine the number of binding sites of ligand, *n*, and the intrinsic association constant for the ligand, *K*, provided there is no interaction of binding sites.

When positive cooperativity or negative cooperativity of binding sites is operating, the Scatchard plot (Fig. 7-9) is no longer linear. A positive homotropic interaction, or positive cooperativity, results in a line which is concave downward (Curve B in Fig. 7-9). While the binding constants in operation may not be easily evaluated from this curved line, the total number of ligand binding sites, *n*, may be deduced from the intercept of the vertical axis. A negative homotropic interaction, or negative cooperativity, results in a line which is concave upward (Curve C in Fig. 7-9). In some instances

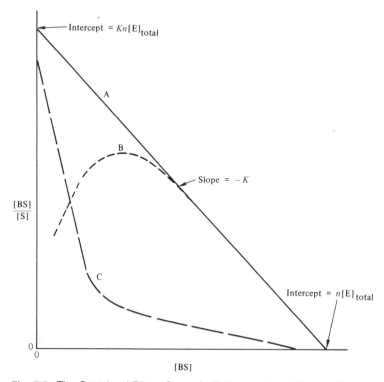

Fig. 7-9. The Scatchard Plot. Curve A: Pattern obtained for noninteracting binding sites. Curve B: Pattern obtained for positive interaction of sites, or positive cooperativity of ligand binding. Curve C: Pattern obtained for negative interaction of sites, or negative cooperativity of ligand binding.

this curve may be approximated by two or more linear segments. The segments may be treated independently to evaluate the number of ligand binding sites, n, involved in each, and the association constants, K, which pertain to them. While evaluation of constants from these curved lines may be hazardous, useful quantitative data may be obtained if this treatment is approached with caution.

In summary then, a linear Scatchard plot is indicative of noninteracting sites in the enzyme under study. A line which is concave downward indicates positive cooperativity of ligand binding, which is consistent with the Hill, MWC-concerted, KNF-sequential,

175

Frieden, and Bernhard models. A line which is concave upward is indicative of negative cooperativity; this result is only consistent with the KNF-sequential and Bernhard asymmetric dimer models. This discussion of methods for experimental differentiation among models must include a final note of caution. The conclusion of negative cooperativity from any of the four tests listed above is considered as being indicative of the operation of the KNF-sequential or the Bernhard equilibrium model. This conclusion presupposes the presence of only one homogeneous enzyme. If two enzymes are present which perform the same function, and yet have different kinetic or substrate binding properties, these tests would yield results indistinguishable from true negative cooperativity. Consequently, the possibility of the presence of more than one enzyme must always be ruled out before one can deduce that negative cooperativity is actually being observed.

The Hysteretic Enzyme Concept

All of the models of enzyme-catalyzed reactions considered to this point are based on initial velocity kinetic data. They presume that, upon mixing enzyme and substrate, there is an instantaneous rate of substrate disappearance by a sequence of enzyme-catalyzed steps. However, numerous enzymes respond slowly to the addition of substrate or rapid changes in ligand concentration. For example, phosphorylase a of rabbit muscle responds to addition of its substrate glycogen with an increase in velocity which takes a period of minutes. Similarly, glyceraldehyde-3-phosphate dehydrogenase of yeast responds to addition of one of its substrates, diphosphopyridine nucleotide (DPN), with a slow increase in velocity. Threonine deaminase requires minutes to respond to the addition of its inhibitor isoleucine. Frieden has termed these *hysteretic enzymes* in analogy to the term *hysteresis,* which is used in physics to define the lag time exhibited by a body in reacting to outside forces. A typical *hysteretic effect* on the disappearance of substrate as a function of time in an enzymatic reaction is shown in Fig. 7-10.

Frieden proposed several possible explanations for hysteresis. The two most important of these are a slow ligand-induced isomerization and a ligand-induced dissociation or association of

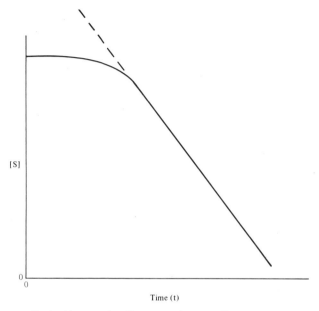

Fig. 7-10. Typical hysteretic effect on substrate disappearance.

oligomeric enzyme. These possible models are generalized in very simplified form as,

$$E + S \rightleftharpoons ES \rightleftharpoons E'S \rightarrow E' + Product, \tag{42}$$

where ES is the less active and E'S the more active conformation or state of association of the enzyme–substrate complex. If the rate of conversion of ES to E'S is slower than the conversion of E'S to E' and product, the enzyme shows a time-dependent change in activity. In other words, there is a lag period in substrate disappearance, as seen in Fig. 7-10. If, on the other hand, the conversion of ES and E'S is rapid relative to decomposition to product, there is no lag period. Hysteretic behavior can be discussed in terms of the MWC-concerted model, the KNF-sequential model, the Frieden association–dissociation, or the Bernhard asymmetric dimer models. The hysteretic response only indicates that a measurable amount of time is required to convert one form of the enzyme to another.

177

While the hysteretic effect has been observed in in vitro experiments, its possible regulatory role in vivo should be considered. It is possible that a slow response of catalytic activity to ligand concentration would insure a constant flow of metabolite down a pathway. A hysteretic response of an enzyme would resist a momentary flux in substrate or effector concentration, but would allow response to a continued change.

REFERENCES

Jean-Pierre Changeux, "The Control of Biochemical Reactions," *Scientific American* **212**, No. 4, 36–45 (1965).

Henry R. Mahler and Eugene H. Cordes, *Biological Chemistry,* 2nd edition, Chapter 6, Harper and Row, New York, 1971.

Keith J. Laidler and Peter S. Bunting, *The Chemical Kinetics of Enzyme Action,* 2nd edition, Chapter 11, Clarendon Press, Oxford, New York, 1973.

D. E. Koshland Jr., "The Molecular Basis of Enzyme Regulation," in *The Enzymes* Paul D. Boyer, Ed., 3rd edition, Vol. 1, Chapter 7, Academic Press, New York, 1970.

This chapter, more than any other, describes recent approaches and developments. Therefore, it is appropriate to summarize the most important original literature.

V. A. Hill, "The Possible Effects of the Aggregation of the Molecules of Hemoglobin on Its Dissociation Curves," *J. Physiol. (London)* **40**, iv-viii (1910).

Daniel E. Atkinson, James A. Hathway, and Eddie C. Smith, "Kinetics of Regulatory Enzymes," *J. Biol. Chem.* **240**, 2682–2690 (1965).

A. Cornish-Bowden and D. E. Koshland Jr., "Diagnostic Use of the Hill Plots," *J. Mol. Biol.* **95**, 201–212 (1975).

Jacques Monod, Jeffries Wyman, and Jean-Pierre Changeux, "On the Nature of Allosteric Transitions: A Plausible Model," *J. Mol. Biol.* **12**, 88–118 (1965).

D. E. Koshland Jr., G. Nemethy, and D. Filmer, "Comparison of Experimental Binding Data and Theoretical Models in Proteins Containing Subunits," *Biochemistry* **5**, 365–385 (1966).

Carl Frieden, "Treatment of Enzyme Kinetic Data II. The Multisite Case: Comparison of Allosteric Models and a Possible New Mechanism," *J. Biol. Chem.* **242**, 4045–4052 (1967).

F. Seydoux, O. P. Malhotra, and S. A. Bernhard, "Half-Site Reactivity," *Critical Reviews in Biochemistry* **2**, 227–257 (1974).

B. R. Rabin, "Co-operative Effects in Enzyme Catalysis: A Possible Kinetic Model Based on Substrate Induced Conformation Isomerization," *Biochem. J.,* 22C–23C (1967).

W. Ferdinand, "The Interpretation of Non-Hyperbolic Rate Curves for Two-Substrate Enzymes," *Biochem. J.* **98**, 278–283 (1966).

George Scatchard, "The Attractions of Proteins for Small Molecules and Ions," *Ann. N.Y. Acad. Sci.* **51**, 660–672 (1949).

Carl Frieden, "Kinetic Aspects of Regulation of Metabolic Processes. The Hysteretic Enzyme Concept," *J. Biol. Chem.* **245**, 5788–5799 (1970).

PROBLEMS

1. Pyruvate kinase is an enzyme of the glycolytic pathway which catalyzes the substrate-level synthesis of ATP with simultaneous conversion of phosphoenolpyruvate to pyruvate:

$$
\begin{array}{c}
O^{\ominus} \\
| \\
{}^{\ominus}O\!-\!P\!=\!O \\
| \\
O \\
| \\
H_2C\!=\!C\!-\!CO_2^{\ominus}
\end{array}
+ ADP \rightarrow CH_3\!-\!\overset{\overset{\displaystyle O}{\|}}{C}\!-\!CO_2^{\ominus} + ATP.
$$

The enzyme from chicken liver has a molecular weight of approximately 235,000, and it contains four subunits. It is activated and inhibited in vitro by a wide range of physiologically occurring substances which may also modify its catalytic activity in vivo. The figure on p. 180 shows how the velocity of the pyruvate kinase-catalyzed reaction, *v,* varies as a function of phosphoenolpyruvate concentration in the absence of and in the presence of a number of these substances; the concentration of ADP is held high and constant. (Data for this and the following problems were generously supplied by Dr. Kenneth Ibsen.)

a. What does the shape of the curve obtained in the absence of any modifier indicate about the mechanism of the enzymatic reaction?

b. Identify the modifier as activators or inhibitors; also indicate if they are *K*-type, *V*-type, or simultaneously *K* and *V*-types.

2. The velocities of the pyruvate kinase reaction obtained at a number of concentrations of phosphoenolpyruvate, at a constant con-

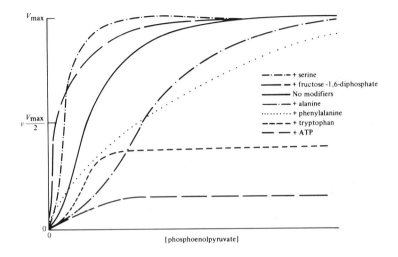

centration of ADP, and in the absence of modifiers, are sum-marized in the accompanying table.

[Phosphoenolpyruvate] (mM)	v (arbitrary units)
0.195	0.006
0.0325	0.012
0.065	0.031
0.195	0.094
0.0325	0.114
∞	0.130

a. What is the shape of the Lineweaver-Burk plot obtained from this data?

b. What is the slope of the line of a Hill plot derived from this data? What is the apparent value of K'?

c. With which equilibrium model of regulatory action by an enzyme is this data consistent?

3. The velocities of the pyruvate kinase reaction, obtained at several concentrations of phosphoenolpyruvate at a constant concentration of ATP and in the presence of fructose-1,6-diphosphate, are summarized in the accompanying table.

180

[Phosphoenolpyruvate] (mM)	v (arbitrary units)
0.0065	0.055
0.013	0.078
0.0195	0.090
0.0325	0.104
∞	0.133

a. What is the shape of the Lineweaver-Burk plot obtained from this data?

b. What is the slope of the line of a Hill plot derived from this data? What is the apparent value of K'?

c. Which equilibrium model of regulatory action is consistent with the velocity data obtained in the presence of fructose-1,6-diphosphate?

4. The velocities of the pyruvate kinase-catalyzed reaction determined at a number of concentrations of phosphoenolpyruvate at a constant concentration of ADP and in the presence of phenylalanine are given in the accompanying table.

[Phosphoenolpyruvate] (mM)	v (arbitrary units)
0.0195	0.018
0.0325	0.024
0.065	0.032
0.13	0.039
0.195	0.055
0.325	0.057
3.25	0.094
∞	0.130

a. What is the shape of the Lineweaver-Burk plot obtained from this data?

b. What is the slope of the line of a Hill plot obtained from this data, and what is the value of K'?

c. Which equilibrium model of regulatory action appears to be functioning in the presence of phenylalanine?

5. The enzyme BSase has a sigmoid-shaped dependence of

181

velocity on substrate concentration. However, the substrate binding data yields a linear Scatchard plot.

a. If BSase has only one substrate, which model or models is consistent with the data?

b. If there are two substrates, which models are consistent with the data?

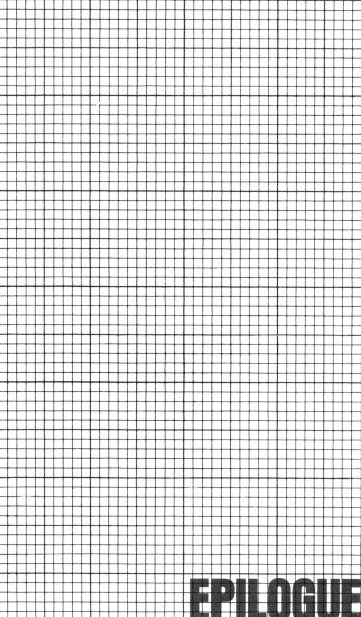

EPILOGUE

The study of kinetics can, and has, contributed much to the under-standing of the mechanisms of chemical and enzyme-catalyzed reactions; however, kinetics like any other discipline has its strengths and limitations. The purpose of these concluding para-graphs is to summarize the applications and limitations of kinetic analysis of chemical and enzymatic reactions. At the outset of this text, I stated that kinetic analyses of reactions are carried out in two identifiable steps, the collection of experimental data in the laboratory and the development of a mathematical treatment or theory which can be used to explain the data. There are possible problems in both of these steps which can arise and complicate the kinetic analysis. These should be recognized.

It should be obvious to all that an interpretation of kinetic data can be no better than the data itself. An occasional data point which falls off the theoretical line of a graphical treatment may be held suspect as the possible result of a random error. If experi-mental data points are scattered around this theoretical line, or if the numerical data have a low degree of reproducibility, a re-examination of the methods being employed is in order. While it is possible to overinterpret data having a low level of reproducibility, it is also possible to underinterpret valid data. The experimenter should be wary of discarding those data points which consistently and inconveniently insist on falling off the theoretical line in the graphical analysis. They may not fall on the theoretical line be-cause they are not on the line. For example, the hydrolysis of p-nitrophenyl acetate by chymotrypsin exhibits a burst-turnover rate of production of p-nitrophenolate (Chapter 5). It would have been very easy to discard the few points which indicated the initial rapid burst of p-nitrophenolate production and build an interpreta-tion on only the steady-state formation of product. Consequently, their implication of a covalent enzyme–substrate intermediate in the reaction mechanism would have been missed. Similarly, by discarding the initial data points for many enzymatic reactions, many investigators missed the very important phenomenon of hysteresis (Chapter 7). Also, many enzymatic reactions produce Lineweaver-Burk plots which are slightly concave upward or down-ward. While they may reflect inaccurate data, they may also de-mand a theory more sophisticated than simple Michaelis-Menten complex formation. They may imply positive or negative coopera-tivity in the binding of substrate to the enzyme (Chapter 7). In summary, if all of the data do not fit the theory, it is possible that the theory should be discarded rather than the data, and be replaced by a better theory.

The theoretical analysis of kinetic data is also subject to ambiguity. The possibility of deriving more than one rate law which can accommodate the experimental data has been stated several times in this text and bears repeating. The possibility of postulating kinetically equivalent rate laws occurs when the reactant or reactants exist in equilibrium with another form. The salient point here is that kinetics does not give the stoichiometry of reactants, even though the rate laws include their concentrations. Kinetics gives the stoichiometry of the activated complex. Any equilibrium or series of equilibria which relate the reactants and the transition state may describe the reaction pathway. This fact dictates the method of operation commonly used in the analysis of kinetic data: the identification of all of the possible kinetically equivalent mechanisms of a reaction, and, by various means, the elimination of all but one possibility which by default is concluded to be correct.

Despite these limitations, kinetics has great power. It can be used as a tool in elucidating reaction mechanisms from their general features to their fine details. In the study of enzymes, kinetics has been used to develop concepts ranging from the enzyme–substrate complex through allosteric interactions to the nature of bond-forming and bond-breaking steps. It is the only technique that examines catalysis from a dynamic point of view. Other physical techniques used to study enzymes, for example spectral analysis and X-ray crystallography, examine only static forms of the enzyme. While these other techniques are important, only kinetics observes catalysis on the time scale in which it occurs. Kinetics is and will continue to be the most important tool in the elucidation of mechanisms of chemical and enzyme-catalyzed reactions.

APPENDIX I
GLOSSARY

In the pursuit of economy of words and precision of description, science devises its own language which may seem to be jargon to the layman or novice. While the goal is admirable, the net effect is often to obscure the meaning of terms used to all but those intimately involved in the field. The purpose of this glossary is to clarify and define the common terminology of kinetics. Many of the terms described below are used in this text; many others are found in the literature. The following list is intended as a handy reference source for reading both this text and the literature.

Abscissa. The horizontal coordinate of a point in a plane Cartesian coordinate system, obtained by measuring parallel to the *x* axis from the *y* axis.

Absolute Temperature. Temperature measured on the absolute scale expressed in degrees Kelvin, after Lord Kelvin, its originator; $°K = °C + 273.15$.

Activated Complex (Transition State). A short-lived, unstable species with a structure intermediate between reactant(s) and product(s).

Activator. A substance that increases the velocity of an enzyme-catalyzed reaction.

Active Site. That region on an enzyme's surface within which all enzyme–substrate interactions take place (see catalytic site).

Allostery. The modification of enzymatic activity resulting from attachment of a substrate or modifier at a site other than the catalytic site. This term has been used in the literature with a great lack of precision. It merely describes a possible mechanism of activation or inhibition. It does not state whether a kinetic effect is homotropic or heterotropic. It does not provide an explanation of sigmoid kinetics.

Alternate Product Inhibitor. An inhibitor which could have been a normal product produced at a reasonable rate if an alternate substrate had been used, and which combines only with the enzyme form it would combine with as a substrate in a reverse reaction.

Apoenzyme. The protein component of an enzyme which requires a cofactor to be catalytically active.

Brønsted Acid. A substance that has a tendency to lose a proton.

Brønsted Base. A substance that has a tendency to take on a proton.

Calorie. The quantity of heat needed to raise the temperature of one gram of water from 14.5 to 15.5°C. Common abbreviations: calorie, cal; kilo calorie, kilo cal, or kcal.

Catalyst. A substance that increases the rate of a chemical reaction, but does not appear in the balanced chemical equation.

Catalytic Site. That portion of the enzyme surface where the bond-forming and bond-breaking steps are effected (see active site).

Closed System. A region of space the boundaries of which may not be crossed by matter, but may be crossed by energy.

Coenzyme. A cofactor which is bound to an enzyme by noncovalent bonds and dissociates as a product during each reaction cycle.

Cofactor. A nonprotein substance which may be required for an enzyme to be catalytically active. A cofactor may be a metal ion or an organic molecule.

Competitive Inhibition. A form of inhibition of enzymatic reactions in which the inhibitor increases the apparent K_m of the varied substrate but has no effect on V_{max}. Both inhibitor and substrate are assumed to bind at the same site on the enzyme; hence, they compete for it.

Concerted. Occurring in unison. In a concerted conformational change of an oligomeric enzyme, all subunits change their conformations simultaneously.

Cooperativity. Homotropic interaction; stimulation or inhibition of the interaction of additional molecule(s) of ligand to an enzyme as a result of interaction of the first molecule(s) of the same ligand with the enzyme.

Data. Factual information used as a basis for reasoning; in the context of this book, information obtained from an experiment.

Dead-End Inhibitor. A compound which inhibits an enzymatic reaction by forming a complex with the enzyme, which cannot go on to form product.

Effector (Modulator). A substance which will cause either an increase or a decrease in the velocity of an enzymatic reaction; an activator or an inhibitor.

Electrophile. An "electron-loving" reactant; a substance which is electron poor and may accept an electron pair (i.e., act as a Lewis acid) in a reaction.

Enzyme. A catalyst which is produced within a living organism.

Equilibrium. A state of balance in which the properties of a system do not change and the concentrations of substances within the system do not change.

Feedback Inhibition (End-Product Inhibition). Inhibition of the catalytic activity of the first enzyme of a metabolic pathway by the end-product of that pathway.

General Acid. A weak acid which may donate a proton in a catalytic mechanism.

General Base. A weak base which may accept a proton in a catalytic mechanism.

Half-Life ($t_{1/2}$). That period of time required for half of the reactant(s) to be transformed in a reaction.

Half of the Sites Reactivity. For an enzyme having multiple subunits, catalysis by or reaction of only half of the subunits of active sites. This term is most appropriate when used to describe extreme forms of negative cooperativity.

Heterotropic Effector. A substance which by interacting with an enzyme will inhibit or stimulate the interaction of a different substance with that enzyme.

Holoenzyme. An intact enzyme–cofactor complex.

Homotropic Effector. A substance which by interacting with an enzyme will stimulate or inhibit the interaction of additional molecule(s) of the same substance with that enzyme.

Hysteresis. The occurrence of a lag time in the response of a body or enzyme in reacting to an outside force or stimulus.

Hysteretic Enzyme. An enzyme which responds slowly in some kinetic characteristic to a rapid change in concentration of substrate or modifier.

Inactivation. The process of blocking an enzymatic reaction in an irreversible manner.

Inactivator. A substance which will decrease the velocity of an enzyme-catalyzed reaction or abolish the reaction, generally by forming a covalent complex with the enzyme.

Inhibition. The process of blocking an enzymatic reaction, generally in a reversible manner.

Inhibitor. A substance which will decrease the velocity of an enzyme-catalyzed reaction, generally by forming a noncovalent complex with the enzyme.

Ion Product of Water. The product of hydrogen ion and hydroxide ion concentrations. Under standard conditions $K_w = [H^+][OH^-] = 10^{-14}M^2$.

Isozymes (Isoenzymes). Enzymes which perform the same catalytic function, but exist in different forms which may or may not show differences in kinetic properties, behavior on electrophoresis, etc.

K-System. An enzymatic reaction in which an activator or inhibitor will either increase or decrease the apparent affinity of enzyme for substrate; i.e., it will either decrease or increase the apparent value of the dissociation constant, K, of the substrate.

Kinetics. The study of the velocities of chemical reactions for the purpose of deducing reaction mechanisms.

Lewis Acid. A substance which may accept a pair of electrons.

Lewis Base. A substance which may donate a pair of electrons.

Ligand. A general term used to describe one molecule which binds to another. In the context of enzymology, a substrate, activator, or inhibitor which brings about a regulatory effect by binding to an enzyme.

Mechanism. All of the individual processes or steps through which reactant(s) must pass to form the product(s) of a reaction.

Modulator (Effector). A substance which will cause either an increase or a decrease in the velocity of an enzyme reaction; an activator or an inhibitor.

Molecularity. The number of molecules of reactant which collide during a reaction process.

Monomer. The fully dissociated subunit of an oligomer or polymer.

Negative Cooperativity. Negative homotropic interaction; inhibition of the interaction of additional molecule(s) of ligand by interaction of the first molecule(s) of ligand with an enzyme. This term is occasionally used interchangeably with *half of the sites reactivity*, which is an extreme form of negative cooperativity.

Noncompetitive Inhibition. A form of inhibition of enzymatic reactions in which the inhibitor increases both the vertical intercept and the slope of a double reciprocal plot.

Nucleophile. A nucleophilic or "nucleus-loving" reactant; a substance which is electron rich and may donate an electron pair (i.e., act as a Lewis base) in a reaction.

Oligomer. An aggregated structure composed of a limited number of subunit structures.

Open System. A region of space the boundaries of which may be crossed by matter and energy.

Order (of a Chemical Reaction). The sum of the exponents of the concentrations of reactants in the rate law for a chemical reaction.

Ordered Mechanism. A multiple substrate enzyme-catalyzed reaction which proceeds via a central complex containing all reactants, and in which substrates add or products are released in an ordered manner.

Ordinate. The vertical coordinate of a point in a plane Cartesian coordinate system, obtained by measuring parallel to the y axis.

Ping Pong. Cleland's term for a double displacement enzyme mechanism.

Polymer. An aggregated structure composed of similar, if not identical, subunits.

Positive Cooperativity. Positive homotropic interaction; stimulation of the interaction of additional molecule(s) of ligand by interaction of the first molecule(s) of ligand with an enzyme.

Product Inhibitor. A compound that is a normal product of an enzymatic reaction and inhibits by combining only with the enzyme form it would react with as a substrate in the reverse reaction.

Prosthetic Group. A cofactor which is covalently bound to an enzyme.

Pseudo First-Order Reaction. A reaction which takes place at a rate which obeys the theoretical rate law for a first-order reaction. The conditions under which the reaction is studied have generally been selected so as to simplify the kinetic analysis to that of a first-order reaction. For example, in a two-reactant reaction, the concentration of one reactant may be held constant by keeping its concentration in vast excess over that of the second. Thus, the concentration of only one reactant will change with time; and the reaction will appear to be first-order.

Random Mechanism. A multiple substrate enzyme-catalyzed reaction which proceeds via a central complex containing all reactants, and in which substrates may attach to the enzyme in any order and products may be released in any order.

Rate Law. A theoretical expression that describes the rate of a reaction as a function of reactant concentration(s). For example, the rate law for the conversion of A to P by a first-order reaction is $-d[A]/dt = k[A]$.

Relaxed (R) State. The conformation of an oligomeric protein which is altered so as to facilitate binding of ligand. In the R state, the dissociation constant(s) for ligand is small relative to that of the tight (T) state.

Saturation. The degree to which ligand binding sites of an enzyme are occupied. At a concentration of substrate or ligand sufficiently high to fill all binding sites, the enzyme is said to be saturated.

Sequential. Occurring in a continuous or connected series. In a sequential conformational change of an oligomeric enzyme, each subunit changes its conformation in an individual, discrete step.

Sequential Mechanism. A multiple substrate enzyme-catalyzed reaction in which reaction proceeds via a central complex containing all reactants.

Sigmoid. Curved in two directions like the letter S.

Specific Acid. An acidic species related to the solvent which may act as a catalyst. In water, specific acid is hydrogen ion.

Specific Base. A basic species related to the solvent which may act as a catalyst. In water, specific base is hydroxide ion.

Standard State. The standard or reference conditions used when measuring thermodynamic activation parameters: unit activity of reactants (generally approximated as one molar) and 25°C.

Steady State. The condition in which rate of formation of a substance is identical to the rate of its decomposition; hence, the concentration of that substance remains constant with time. The "steady-state assumption" is of use in the derivation of kinetic rate laws such as the Michaelis-Menten equation (Chapter 5).

Substrate. The reactant in a chemical or enzyme-catalyzed reaction.

Surroundings. In thermodynamics, all of the universe which is excluded from a defined system.

System. In thermodynamics, the matter within a defined region.

Ternary Complex. A complex of three substances; for example, an enzyme and two different substrate molecules.

Tetramer. A structure composed of four subunits.

Theory. A plausible general principle offered to explain phenomena; a hypothesis assumed for the sake of argument or investigation; in the context of this book, a mathematical treatment used to quantitatively describe experimental data.

Thermodynamics. The science which studies energy changes in physical and chemical processes.

Tight (T) State. The conformation of an oligomeric protein which is altered so as to restrict the binding of ligand. In the T state, the dissociation constant(s) for ligand is large relative to that of the relaxed (R) state.

Transition State (Activated Complex). A short-lived, unstable species with a structure intermediate between reactant(s) and product(s).

Turnover Number. The maximum number of substrate molecules converted to product per unit time by a single enzyme molecule. It is a first-order rate constant with units of time^{-1}.

Uncompetitive Inhibition. A form of inhibition of enzymatic reaction in which the inhibitor increases the vertical intercept of a double reciprocal plot, but has no effect on its slope.

V-System. An enzymatic reaction in which an activator or inhibitor will either increase or decrease the maximum attainable velocity (V_{max}) of the reaction.

Zymogen. An inactive precursor of an enzyme which is activated by cleavage of one or a few specific peptide bonds.

APPENDIX II
CONSTANTS
AND COMMON
SYMBOLS

This appendix is a list of symbols used in this text. They are also the most common abbreviations used in the literature. Therefore, it is the author's hope that this list may serve not only as a guide to this text, but may also be useful in reading from other sources.

a_H Hydrogen ion activity as measured by a pH meter. Except at extreme concentrations of acid or base, $a_H \approx [H^+]$.

A,B,C. . .   The substrates in a multiple-substrate enzymatic reaction.

α Brønsted coefficient; a measure of the sensitivity of a reaction to catalysis by a series of general acids.

β Brønsted coefficient; a measure of the sensitivity of a reaction to catalysis by a series of general bases or nucleophiles.

e The base of natural logarithms; 2.71828.

ϵ Energy of vibration.

e.u.   Entropy units, cal deg^{-1} $mole^{-1}$.

E Energy.

E   Enzyme.

E'   A modified form of an enzyme which is produced in a double displacement, or Ping Pong, enzymatic mechanism.

E_a The Arrhenius energy of activation.

E_r The "relaxed" conformation of an enzyme which has a greater affinity for ligand than does the "tight" or E_t conformation.

E_t The "tight" conformation of an enzyme which has a lesser affinity for ligand than does the "relaxed" conformation.

F (G is used in the newer literature) Free energy.

G (F is used in the older terminology) Free energy.

ΔG^0 Difference in free energy levels of reactant(s) and product(s) under standard-state conditions.

$\Delta G^{\ddagger}$ Free energy of activation; the difference in free energy levels of reactant(s) and transition state.

h Planck's constant; 1.584×10^{-34} cal·sec.

H Enthalpy or heat content.

ΔH^0 Difference in enthalpy levels of reactant(s) and product(s) under standard state conditions.

197

$\Delta H^{\ddagger}$ Enthalpy of activation; the difference in enthalpy levels of reactant(s) and transition state.

I   Inhibitor of an enzymatic reaction.

k Commonly used symbol for a rate constant.

$k^{\ddagger}$ Rate constant for breakdown of an activated complex.

k_B Boltzmann's constant; 3.298×10^{-24} cal·deg^{-1}; 1.381×10^{-23} joules·deg^{-1}.

k_B Second-order rate constant for a general base or nucleophile-catalyzed reaction.

k_{BH} Second-order rate constant for a general acid-catalyzed reaction.

k_H The rate constant of a reaction catalyzed by hydrogen ion. This term is derived from experimental data; it is a second-order rate constant where $k_{obsd} = k_H[H^+]$.

k_{obsd} A rate constant which has been observed for a chemical reaction under a defined set of conditions. It is generally understood to be a complex constant equal to the actual rate constant of the reaction multiplied by terms which may include, among others, concentration of reactant(s) and dissociation constants.

k_{OH} The rate constant of a reaction catalyzed by hydroxide ion. This term is derived from experimental data; it is a second-order rate constant where $k_{obsd} = k_{OH}[OH^-]$.

k_W The rate constant of a reaction catalyzed by water or involving water as a reactant. This term is derived from experimental data; it is a second-order rate constant where $k_{obsd} = k_W[H_2O]$.

K Commonly used symbol for an equilibrium constant.

K' In the Hill model, the constant for the breakdown of ES_n; i.e., $K' = (k_2 + k_3)/k_1$. It is analogous to the Michaelis-Menten constant.

$K^{\ddagger}$ Equilibrium constant for formation of an activated complex.

K_a Dissociation constant for a Brønsted acid.

K_a In Cleland's terminology describing multiple substrate enzymatic reactions, the Michaelis-Menten constant of substrate A.

K_{app} A dissociation constant which is deduced from kinetic data. It is termed an apparent dissociation constant because it may include other equilibrium terms in addition to the dissociation.

K_b In Cleland's terminology describing multiple substrate enzymatic reactions, the Michaelis-Menten constant for substrate B.

K_d Dissociation constant.

K_e The kinetically apparent dissociation constant of a group of an uncomplexed enzyme.

K_{es} The kinetically apparent dissociation constant of a group of an enzyme–substrate complex.

K_I Dissociation constant of an enzyme–inhibitor complex.

K_{ia} In Cleland's terminology describing multiple substrate enzymatic reactions, the dissociation constant of substrate A.

K_m The Michaelis-Menten constant. It is equal to the sum of the rate constants for the decomposition of enzyme–substrate complex to product and to free reactant, divided by the rate constant for formation of enzyme–substrate complex. Experimentally it is that concentration of substrate at which the observed velocity, v, is equal to half the maximum attainable velocity, $V_{max}/2$, i.e., $K_m = [S]$ when $v = V_{max}/2$.

K_m' A pH-dependent Michaelis-Menten constant.

$K_m{}^A$ In the terminology of the Enzyme Commission describing multiple substrate enzymatic reactions, the Michaelis-Menten constant of substrate A.

$K_m{}^B$ In the terminology of the Enzyme Commission describing multiple substrate enzymatic reactions, the Michaelis-Menten constant of substrate B.

K_r The dissociation constant of substrate from enzyme in its "relaxed" conformational form.

K_s The dissociation constant of an enzyme–substrate complex.

$K_s{}^A$ In the terminology of the Enzyme Commission describing multiple substrate enzymatic reactions, the dissociation constant of substrate A.

K_t The dissociation constant of substrate from enzyme in its "tight" conformational form.

K_W The dissociation constant of water; equal to the product of the concentrations of hydrogen ion and hydroxide ion.

κ Transmission coefficient; the probability that an activated complex will decompose to products. It is generally assumed to equal 1.0.

199

L The allosteric equilibrium constant. It is equal to the ratio of the concentrations of protein in the relaxed and tight states; i.e. $L = [E_r]/[E_t]$.

n The coefficient of the Hill equation which is commonly viewed as an index of the degree of cooperativity of ligand binding.

n The number of binding sites for ligands per enzyme molecule.

ν Vibrational frequency of an activated complex which leads to its decomposition.

P Pressure.

P,Q,R. . .   The products in a multiple substrate enzymatic reaction.

pH   The commonly used measure of acidity. $pH = -\log[H^+] = \log 1/[H^+]$.

pK_a A measure of the strength of a substance as a Brønsted acid as a function of its dissociation constant. $pK_a = -\log K_a = \log 1/K_a$.

pK_{app} The negative logarithm of an apparent dissociation constant deduced from kinetic data.

Q Heat.

R Gas constant; 1.987 cal·deg^{-1} mole^{-1}; 8.314 joules·deg^{-1} mole^{-1}.

ρ The Hammett reaction constant. It is a measure of the susceptibility of the rate of a reaction to polar effects.

S Entropy.

S   Substrate. The reactant in an enzymatic reaction.

ΔS^0 Difference in entropy levels of reactant(s) and product(s) at standard-state conditions.

$\Delta S^{\ddagger}$ Entropy of activation; the difference in entropy levels of reactant(s) and transition state.

σ The Hammett substitutent constant. $\sigma = \log(K/K_0)$ where K is the dissociation constant of a substituted benzoic acid, and K_0 is the dissociation constant of benzoic acid.

t Time.

$t_{1/2}$ Half-life.

T Temperature, usually expressed on the absolute scale in degrees Kelvin; $^{\circ}K = {^{\circ}C} + 273.15$.

v Observed velocity or observed rate of a reaction; used most often for enzyme-catalyzed reactions.

200

V Volume.

$V(V_{max})$ The maximum velocity which may be attained by an enzyme-catalyzed reaction.

V' In a multiple substrate enzyme-catalyzed reaction, the maximum velocity which may be attained by varying one substrate while holding the other(s) constant.

V_1 In Cleland's terminology, the maximum velocity of an enzyme-catalyzed reaction in the forward direction.

V_2 In Cleland's terminology, the maximum velocity of an enzyme-catalyzed reaction in the reverse direction.

V'_m A pH-dependent maximum velocity for an enzymatic reaction.

W Work.

X Intermediate in a reaction scheme.

Y The saturation function, it defines the degree of ligand binding to a protein or the fraction of the maximal reaction velocity.

APPENDIX III
LOGARITHMIC
TABLES

Logarithms

N	0	1	2	3	4	5	6	7	8	9
10	0000	0043	0086	0128	0170	0212	0253	0294	0334	0374
11	0414	0453	0492	0531	0569	0607	0645	0682	0719	0755
12	0792	0828	0864	0899	0934	0969	1004	1038	1072	1106
13	1139	1173	1206	1239	1271	1303	1335	1367	1399	1430
14	1461	1492	1523	1553	1584	1614	1644	1673	1703	1732
15	1761	1790	1818	1847	1875	1903	1931	1959	1987	2014
16	2041	2068	2095	2122	2148	2175	2201	2227	2253	2279
17	2304	2330	2355	2380	2405	2430	2455	2480	2504	2529
18	2553	2577	2601	2625	2648	2672	2695	2718	2742	2765
19	2788	2810	2833	2856	2878	2900	2923	2945	2967	2989
20	3010	3032	3054	3075	3096	3118	3139	3160	3181	3201
21	3222	3243	3263	3284	3304	3324	3345	3365	3385	3404
22	3424	3444	3464	3483	3502	3522	3541	3560	3579	3598
23	3617	3636	3655	3674	3692	3711	3729	3747	3766	3784
24	3802	3820	3838	3856	3874	3892	3909	3927	3945	3962
25	3979	3997	4014	4031	4048	4065	4082	4099	4116	4133
26	4150	4166	4183	4200	4216	4232	4249	4265	4281	4298
27	4314	4330	4346	4362	4378	4393	4409	4425	4440	4456
28	4472	4487	4502	4518	4533	4548	4564	4579	4594	4609
29	4624	4639	4654	4669	4683	4698	4713	4728	4742	4757
30	4771	4786	4800	4814	4829	4843	4857	4871	4886	4900
31	4914	4928	4942	4955	4969	4983	4997	5011	5024	5038
32	5051	5065	5079	5092	5105	5119	5132	5145	5159	5172
33	5185	5198	5211	5224	5237	5250	5263	5276	5289	5302
34	5315	5328	5340	5353	5366	5378	5391	5403	5416	5428
35	5441	5453	5465	5478	5490	5502	5514	5527	5539	5551
36	5563	5575	5587	5599	5611	5623	5635	5647	5658	5670
37	5682	5694	5705	5717	5729	5740	5752	5763	5775	5786
38	5798	5809	5821	5832	5843	5855	5866	5877	5888	5899
39	5911	5922	5933	5944	5955	5966	5977	5988	5999	6010
40	6021	6031	6042	6053	6064	6075	6085	6096	6107	6117
41	6128	6138	6149	6160	6170	6180	6191	6201	6212	6222
42	6232	6243	6253	6263	6274	6284	6294	6304	6314	6325
43	6335	6345	6355	6365	6375	6385	6395	6405	6415	6425
44	6435	6444	6454	6464	6474	6484	6493	6503	6513	6522
45	6532	6542	6551	6561	6571	6580	6590	6599	6609	6618
46	6628	6637	6646	6656	6665	6675	6684	6693	6702	6712
47	6721	6730	6739	6749	6758	6767	6776	6785	6794	6803
48	6812	6821	6830	6839	6848	6857	6866	6875	6884	6893
49	6902	6911	6920	6928	6937	6946	6955	6964	6972	6981
50	6990	6998	7007	7016	7024	7033	7042	7050	7059	7067
51	7076	7084	7093	7101	7110	7118	7126	7135	7143	7152
52	7160	7168	7177	7185	7193	7202	7210	7218	7226	7235
53	7243	7251	7259	7267	7275	7284	7292	7300	7308	7316
54	7324	7332	7340	7348	7356	7364	7372	7380	7388	7396
N	0	1	2	3	4	5	6	7	8	9

N	0	1	2	3	4	5	6	7	8	9
55	7404	7412	7419	7427	7435	7443	7451	7459	7466	7474
56	7482	7490	7497	7505	7513	7520	7528	7536	7543	7551
57	7559	7566	7574	7582	7589	7597	7604	7612	7619	7627
58	7634	7642	7649	7657	7664	7672	7679	7686	7694	7701
59	7709	7716	7723	7731	7738	7745	7752	7760	7767	7774
60	7782	7789	7796	7803	7810	7818	7825	7832	7839	7846
61	7853	7860 ·	7868	7875	7882	7889	7896	7903	7910	7917
62	7924	7931	7938	7945	7952	7959	7966	7973	7980	7987
63	7993	8000	8007	8014	8021	8028	8035	8041	8048	8055
64	8062	8069	8075	8082	8089	8096	8102	8109	8116	8122
65	8129	8136	8142	8149	8156	8162	8169	8176	8182	8189
66	8195	8202	8209	8215	8222	8228	8235	8241	8248	8254
67	8261	8267	8274	8280	8287	8293	8299	8306	8312	8319
68	8325	8331	8338	8344	8351	8357	8363	8370	8376	8382
69	8388	8395	8401	8407	8414	8420	8426	8432	8439	8445
70	8451	8457	8463	8470	8476	8482	8488	8494	8500	8506
71	8513	8519	8525	8531	8537	8543	8549	8555	8561	8567
72	8573	8579	8585	8591	8597	8603	8609	8615	8621	8627
73	8633	8639	8645	8651	8657	8663	8669	8675	8681	8686
74	8692	8698	8704	8710	8716	8722	8727	8733	8739	8745
75	8751	8756	8762	8768	8774	8779	8785	8791	8797	8802
76	8808	8814	8820	8825	8831	8837	8842	8848	8854	8859
77	8865	8871	8876	8882	8887	8893	8899	8904	8910	8915
78	8921	8927	8932	8938	8943	8949	8954	8960	8965	8971
79	8976	8982	8987	8993	8998	9004	9009	9015	9020	9025
80	9031	9036	9042	9047	9053	9058 '	9063	9069	9074	9079
81	9085	9090	9096	9101	9106	9112	9117	9122	9128	9133
82	9138	9143	9149	9154	9159	9165	9170	9175	9180	9186
83	9191	9196	9201	9206	9212	9217	9222	9227	9232	9238
84	9243	9248	9253	9258	9263	9269	9274	9279	9284	9289
85	9294	9299	9304	9309	9315	9320	9325	9330	9335	9340
86	9345	9350	9355	9360	9365	9370	9375	9380	9385	9390
87	9395	9400	9405	9410	9415	9420	9425	9430	9435	9440
88	9445	9450	9455	9460	9465	9469	9474	9479	9484	9489
89	9494	9499	9504	9509	9513	9518	9523	9528	9533	9538
90	9542	9547	9552	9557	9562	9566	9571	9576	9581	9586
91	9590	9595	9600	9605	9609	9614	9619	9624	9628	9633
92	9638	9643	9647	9652	9657	9661	9666	9671	9675	9680
93	9685	9689	9694	9699	9703	9708	9713	9717	9722	9727
94	9731	9736	9741	9745	9750	9754	9759	9763	9768	9773
95	9777	9782	9786	9791	9795	9800	9805	9809	9814	9818
96	9823	9827	9832	9836	9841	9845	9850	9854	9859	9863
97	9868	9872	9877	9881	9886	9890	9894	9899	9903	9908
98	9912	9917	9921	9926	9930	9934	9939	9943	9948	9952
99	9956	9961	9965	9969	9974	9978	9983	9987	9991	9996
N	0	1	2	3	4	5	6	7	8	9

Logarithms of Decimal Fractions

N	0	1	2	3	4	5	6	7	8	9
.10	−1.000	−.9957	−.9914	−.9872	−.9830	−.9788	−.9747	−.9706	−.9666	−.9626
.11	−.9586	−.9547	−.9508	−.9469	−.9431	−.9393	−.9355	−.9318	−.9281	−.9245
.12	−.9208	−.9172	−.9136	−.9101	−.9066	−.9031	−.8996	−.8962	−.8928	−.8894
.13	−.8861	−.8827	−.8794	−.8761	−.8729	−.8697	−.8665	−.8633	−.8601	−.8570
.14	−.8539	−.8508	−.8477	−.8447	−.8416	−.8386	−.8356	−.8327	.8297	−.8268
.15	−.8239	−.8210	−.8182	−.8153	−.8125	−.8097	−.8069	−.8041	−.8013	−.7986
.16	−.7959	−.7932	−.7905	−.7878	−.7852	−.7825	−.7799	−.7773	−.7747	−.7721
.17	−.7696	−.7670	−.7645	−.7620	−.7595	−.7570	−.7545	−.7520	−.7496	−.7471
.18	−.7447	−.7423	−.7399	−.7375	−.7352	−.7328	−.7305	−.7282	−.7258	−.7235
.19	−.7212	−.7190	−.7167	−.7144	−.7122	−.7100	−.7077	−.7055	−.7033	−.7011
.20	−.6990	−.6968	−.6946	−.6925	−.6904	−.6882	−.6861	−.6840	−.6819	−.6799
.21	−.6778	−.6757	−.6737	−.6716	−.6696	−.6676	−.6655	−.6635	−.6615	−.6596
.22	−.6576	−.6556	−.6536	−.6517	−.6498	−.6478	−.6459	−.6440	−.6421	−.6402
.23	−.6383	−.6364	−.6345	−.6326	−.6308	−.6289	−.6271	−.6253	−.6234	−.6216
.24	−.6198	−.6180	−.6162	−.6144	−.6126	.6108	−.6091	−.6073	−.6055	−.6038
.25	−.6021	−.6003	−.5986	−.5969	−.5952	−.5935	−.5918	−.5901	−.5884	−.5867
.26	−.5850	−.5834	−.5817	−.5800	−.5784	−.5768	−.5751	−.5735	−.5719	−.5702
.27	−.5686	−.5670	−.5654	−.5638	−.5622	−.5607	−.5591	−.5575	−.5560	−.5544
.28	−.5528	−.5513	−.5498	−.5482	−.5467	−.5452	−.5436	−.5421	−.5406	−.5391
.29	−.5376	−.5361	−.5346	−.5331	−.5317	−.5302	−.5287	−.5272	−.5258	−.5243
.30	−.5229	−.5214	−.5200	−.5186	−.5171	−.5157	−.5143	−.5129	−.5114	−.5100
.31	−.5086	−.5072	−.5058	−.5045	−.5031	−.5017	−.5003	−.4989	−.4976	−.4962
.32	−.4949	−.4935	−.4921	−.4908	−.4895	−.4881	−.4868	−.4855	−.4841	−.4828
.33	−.4815	−.4802	−.4789	−.4776	−.4763	−.4750	−.4737	−.4724	−.4711	−.4698
.34	−.4685	−.4672	−.4660	−.4647	−.4634	−.4622	−.4609	−.4597	−.4584	−.4572
.35	−.4559	−.4547	−.4535	−.4522	−.4510	−.4498	−.4486	−.4473	−.4461	−.4449
.36	−.4437	−.4425	−.4413	−.4401	−.4389	−.4377	−.4365	−.4353	−.4342	−.4330
.37	−.4318	−.4306	−.4295	−.4283	−.4271	−.4260	−.4248	−.4237	−.4225	−.4214
.38	−.4202	−.4191	−.4179	−.4168	−.4157	−.4145	−.4134	−.4123	−.4112	−.4101
.39	−.4089	−.4078	−.4067	−.4056	−.4045	−.4034	−.4023	−.4012	−.4001	−.3990
.40	−.3979	−.3969	−.3958	−.3947	−.3936	−.3925	−.3915	−.3904	−.3893	−.3883
.41	−.3872	−.3862	−.3851	−.3840	−.3830	−.3820	−.3809	−.3799	−.3788	−.3778
.42	−.3768	−.3757	−.3747	−.3737	−.3726	−.3716	−.3706	−.3696	−.3686	−.3675
.43	−.3665	−.3655	−.3645	−.3635	−.3625	−.3615	−.3605	−.3595	−.3585	−.3575
.44	−.3565	−.3556	−.3546	−.3536	−.3526	−.3516	−.3507	−.3497	−.3487	−.3478
.45	−.3468	−.3458	−.3449	−.3439	−.3429	−.3420	−.3410	−.3401	−.3391	−.3382
.46	−.3372	−.3363	−.3354	−.3344	−.3335	−.3325	−.3316	−.3307	−.3298	−.3288
.47	−.3279	−.3270	−.3261	−.3251	−.3242	−.3233	−.3224	−.3215	−.3206	−.3197
.48	−.3188	−.3179	−.3170	−.3161	−.3152	−.3143	−.3134	−.3125	−.3116	−.3107
.49	−.3098	−.3089	−.3080	−.3072	−.3063	−.3054	−.3045	−.3036	−.3028	−.3019
.50	−.3010	−.3002	−.2993	−.2984	−.2976	−.2967	−.2958	−.2950	−.2941	−.2933
.51	−.2924	−.2916	−.2907	−.2899	−.2890	−.2882	−.2874	−.2865	−.2857	−.2848
.52	−.2840	−.2832	−.2823	−.2815	−.2807	−.2798	−.2790	−.2782	−.2774	−.2765
.53	−.2757	−.2749	−.2741	−.2733	−.2725	−.2716	−.2708	−.2700	−.2692	−.2684
.54	−.2676	−.2668	−.2660	−.2652	−.2644	−.2636	−.2628	−.2620	−.2612	−.2604
N	0	1	2	3	4	5	6	7	8	9

N	0	1	2	3	4	5	6	7	8	9
.55	−.2596	−.2588	−.2581	−.2573	−.2565	−.2557	−.2549	−.2541	−.2534	−.2526
.56	−.2518	−.2510	−.2503	−.2495	−.2487	−.2480	−.2472	−.2464	−.2457	−.2449
.57	−.2441	−.2434	−.2426	−.2418	−.2411	−.2403	−.2396	−.2388	−.2381	−.2373
.58	−.2366	−.2358	−.2351	−.2343	−.2336	−.2328	−.2321	−.2314	−.2306	−.2299
.59	−.2291	−.2284	−.2277	−.2269	−.2262	−.2255	−.2248	−.2240	−.2233	−.2226
.60	−.2218	−.2211	−.2204	−.2197	−.2190	−.2182	−.2175	−.2168	−.2161	−.2154
.61	−.2147	−.2140	−.2132	−.2125	−.2118	−.2111	−.2104	−.2097	−.2090	−.2083
.62	−.2076	−.2069	−.2062	−.2055	−.2048	−.2041	−.2034	−.2027	−.2020	−.2013
.63	−.2007	−.2000	−.1993	−.1986	−.1979	−.1972	−.1965	−.1959	−.1952	−.1945
.64	−.1938	−.1931	−.1925	−.1918	−.1911	−.1904	−.1898	−.1891	−.1884	−.1878
.65	−.1871	−.1864	−.1858	−.1851	−.1844	−.1838	−.1831	−.1824	−.1818	−.1811
.66	−.1805	−.1798	−.1791	−.1785	−.1778	−.1772	−.1765	−.1759	−.1752	−.1746
.67	−.1739	−.1733	−.1726	−.1720	−.1713	−.1707	−.1701	−.1694	−.1688	−.1681
.68	−.1675	−.1669	−.1662	−.1656	−.1649	−.1643	−.1637	−.1630	−.1624	−.1618
.69	−.1612	−.1605	−.1599	−.1593	−.1586	−.1580	−.1574	−.1568	−.1561	−.1555
.70	−.1549	−.1543	−.1537	−.1530	−.1524	−.1518	−.1512	−.1506	−.1500	−.1494
.71	−.1487	−.1481	−.1475	−.1469	−.1463	−.1457	−.1451	−.1445	−.1439	−.1433
.72	−.1427	−.1421	−.1415	−.1409	−.1403	−.1397	−.1391	−.1385	−.1379	−.1373
.73	−.1367	−.1361	−.1355	−.1349	−.1343	−.1337	−.1331	−.1325	−.1319	−.1314
.74	−.1308	−.1302	−.1296	−.1290	−.1284	−.1278	−.1273	−.1267	−.1261	−.1255
.75	−.1249	−.1244	−.1238	−.1232	−.1226	−.1221	−.1215	−.1209	−.1203	−.1198
.76	−.1192	−.1186	−.1180	−.1175	−.1169	−.1163	−.1158	−.1152	−.1146	−.1141
.77	−.1135	−.1129	−.1124	−.1118	−.1113	−.1107	−.1101	−.1096	−.1090	−.1085
.78	−.1079	−.1073	−.1068	−.1062	−.1057	−.1051	−.1046	−.1040	−.1035	−.1029
.79	−.1024	−.1018	−.1013	−.1007	−.1002	−.0996	−.0991	−.0985	−.0980	−.0975
.80	−.0969	−.0964	−.0958	−.0953	−.0947	−.0942	−.0937	−.0931	−.0926	−.0921
.81	−.0915	−.0910	−.0904	−.0899	−.0894	−.0888	−.0883	−.0878	−.0872	−.0867
.82	−.0862	−.0857	−.0851	−.0846	−.0841	−.0835	−.0830	−.0825	−.0820	−.0814
.83	−.0809	−.0804	−.0799	−.0794	−.0788	−.0783	−.0778	−.0773	−.0768	−.0762
.84	−.0757	−.0752	−.0747	−.0742	−.0737	−.0731	−.0726	−.0721	−.0716	−.0711
.85	−.0706	−.0701	−.0696	−.0691	−.0685	−.0680	−.0675	−.0670	−.0665	−.0660
.86	−.0655	−.0650	−.0645	−.0640	−.0635	−.0630	−.0625	−.0620	−.0615	−.0610
.87	−.0605	−.0600	−.0595	−.0590	−.0585	−.0580	−.0575	−.0570	−.0565	−.0560
.88	−.0555	−.0550	−.0545	−.0540	−.0535	−.0531	−.0526	−.0521	−.0516	−.0511
.89	−.0506	−.0501	−.0496	−.0491	−.0487	−.0482	−.0477	−.0472	−.0467	−.0462
.90	−.0458	−.0453	−.0448	−.0443	−.0438	−.0434	−.0429	−.0424	−.0419	−.0414
.91	−.0410	−.0405	−.0400	−.0395	−.0391	−.0386	−.0381	−.0376	−.0372	−.0367
.92	−.0362	−.0357	−.0353	−.0348	−.0343	−.0339	−.0334	−.0329	−.0325	−.0320
.93	−.0315	−.0311	−.0306	−.0301	−.0297	−.0292	−.0287	−.0283	−.0278	−.0273
.94	−.0269	−.0264	−.0259	−.0255	−.0250	−.0246	−.0241	−.0237	−.0232	−.0227
.95	−.0223	−.0218	−.0214	−.0209	−.0205	−.0200	−.0195	−.0191	−.0186	−.0182
.96	−.0177	−.0173	−.0168	−.0164	−.0159	−.0155	−.0150	−.0146	−.0141	−.0137
.97	−.0132	−.0128	−.0123	−.0119	−.0114	−.0110	−.0106	−.0101	−.0097	−.0092
.98	−.0088	−.0083	−.0079	−.0074	−.0070	−.0066	−.0061	−.0057	−.0052	−.0048
.99	−.0044	−.0039	−.0035	−.0031	−.0026	−.0022	−.0017	−.0013	−.0009	−.0004
N	0	1	2	3	4	5	6	7	8	9

Antilogarithms

	0	1	2	3	4	5	6	7	8	9
.00	1000	1002	1005	1007	1009	1012	1014	1016	1019	1021
.01	1023	1026	1028	1030	1033	1035	1038	1040	1042	1045
.02	1047	1050	1052	1054	1057	1059	1062	1064	1067	1069
.03	1072	1074	1076	1079	1081	1084	1086	1089	1091	1094
.04	1096	1099	1102	1104	1107	1109	1112	1114	1117	1119
.05	1122	1125	1127	1130	1132	1135	1138	1140	1143	1146
.06	1148	1151	1153	1156	1159	1161	1164	1167	1169	1172
.07	1175	1178	1180	1183	1186	1189	1191	1194	1197	1199
.08	1202	1205	1208	1211	1213	1216	1219	1222	1225	1227
.09	1230	1233	1236	1239	1242	1245	1247	1250	1253	1256
.10	1259	1262	1265	1268	1271	1274	1276	1279	1282	1285
.11	1288	1291	1294	1297	1300	1303	1306	1309	1312	1315
.12	1318	1321	1324	1327	1330	1334	1337	1340	1343	1346
.13	1349	1352	1355	1358	1361	1365	1368	1371	1374	1377
.14	1380	1384	1387	1390	1393	1396	1400	1403	1406	1409
.15	1413	1416	1419	1422	1426	1429	1432	1435	1439	1442
.16	1445	1449	1452	1455	1459	1462	1466	1469	1472	1476
.17	1479	1483	1486	1489	1493	1496	1500	1503	1507	1510
.18	1514	1517	1521	1524	1528	1531	1535	1538	1542	1545
.19	1549	1552	1556	1560	1563	1567	1570	1574	1578	1581
.20	1585	1589	1592	1596	1600	1603	1607	1611	1614	1618
.21	1622	1626	1629	1633	1637	1641	1644	1648	1652	1656
.22	1660	1663	1667	1671	1675	1679	1683	1687	1690	1694
.23	1698	1702	1706	1710	1714	1718	1722	1726	1730	1734
.24	1738	1742	1746	1750	1754	1758	1762	1766	1770	1774
.25	1778	1782	1786	1791	1795	1799	1803	1807	1811	1816
.26	1820	1824	1828	1832	1837	1841	1845	1849	1854	1858
.27	1862	1866	1871	1875	1879	1884	1888	1892	1897	1901
.28	1905	1910	1914	1919	1923	1928	1932	1936	1941	1945
.29	1950	1954	1959	1963	1968	1972	1977	1982	1986	1991
.30	1995	2000	2004	2009	2014	2018	2023	2028	2032	2037
.31	2042	2046	2051	2056	2061	2065	2070	2075	2080	2084
.32	2089	2094	2099	2104	2109	2113	2118	2123	2128	2133
.33	2138	2143	2148	2153	2158	2163	2168	2173	2178	2183
.34	2188	2193	2198	2203	2208	2213	2218	2223	2228	2234
.35	2239	2244	2249	2254	2259	2265	2270	2275	2280	2286
.36	2291	2296	2301	2307	2312	2317	2323	2328	2333	2339
.37	2344	2350	2355	2360	2366	2371	2377	2382	2388	2393
.38	2399	2404	2410	2415	2421	2427	2432	2438	2443	2449
.39	2455	2460	2466	2472	2477	2483	2489	2495	2500	2506
.40	2512	2518	2523	2529	2535	2541	2547	2553	2559	2564
.41	2570	2576	2582	2588	2594	2600	2606	2612	2618	2624
.42	2630	2636	2642	2649	2655	2661	2667	2673	2679	2685
.43	2692	2698	2704	2710	2716	2723	2729	2735	2742	2748
.44	2754	2761	2767	2773	2780	2786	2793	2799	2805	2812
.45	2818	2825	2831	2838	2844	2851	2858	2864	2871	2877
.46	2884	2891	2897	2904	2911	2917	2924	2931	2938	2944
.47	2951	2958	2965	2972	2979	2985	2992	2999	3006	3013
.48	3020	3027	3034	3041	3048	3055	3062	3069	3076	3083
.49	3090	3097	3105	3112	3119	3126	3133	3141	3148	3155
	0	1	2	3	4	5	6	7	8	9

	0	1	2	3	4	5	6	7	8	9
.50	3162	3170	3177	3184	3192	3199	3206	3214	3221	3228
.51	3236	3243	3251	3258	3266	3273	3281	3289	3296	3304
.52	3311	3319	3327	3334	3342	3350	3357	3365	3373	3381
.53	3388	3396	3404	3412	3420	3428	3436	3443	3451	3459
.54	3467	3475	3483	3491	3499	3508	3516	3524	3532	3540
.55	3548	3556	3565	3573	3581	3589	3597	3606	3614	3622
.56	3631	3639	3648	3656	3664	3673	3681	3690	3698	3707
.57	3715	3724	3733	3741	3750	3758	3767	3776	3784	3793
.58	3802	3811	3819	3828	3837	3846	3855	3864	3873	3882
.59	3890	3899	3908	3917	3926	3936	3945	3954	3963	3972
.60	3981	3990	3999	4009	4018	4027	4036	4046	4055	4064
.61	4074	4083	4093	4102	4111	4121	4130	4140	4150	4159
.62	4169	4178	4188	4198	4207	4217	4227	4236	4246	4256
.63	4266	4276	4285	4295	4305	4315	4325	4335	4345	4355
.64	4365	4375	4385	4395	4406	4416	4426	4436	4446	4457
.65	4467	4477	4487	4498	4508	4519	4529	4539	4550	4560
.66	4571	4581	4592	4603	4613	4624	4634	4645	4656	4667
.67	4677	4688	4699	4710	4721	4732	4742	4753	4764	4775
.68	4786	4797	4808	4819	4831	4842	4853	4864	4875	4887
.69	4898	4909	4920	4932	4943	4955	4966	4977	4989	5000
.70	5012	5023	5035	5047	5058	5070	5082	5093	5105	5117
.71	5129	5140	5152	5164	5176	5188	5200	5212	5224	5236
.72	5248	5260	5272	5284	5297	5309	5321	5333	5346	5358
.73	5370	5383	5395	5408	5420	5433	5445	5458	5470	5483
.74	5495	5508	5521	5534	5546	5559	5572	5585	5598	5610
.75	5623	5636	5649	5662	5675	5689	5702	5715	5728	5741
.76	5754	5768	5781	5794	5808	5821	5834	5848	5861	5875
.77	5888	5902	5916	5929	5943	5957	5970	5984	5998	6012
.78	6026	6039	6053	6067	6081	6095	6109	6124	6138	6152
.79	6166	6180	6194	6209	6223	6237	6252	6266	6281	6295
.80	6310	6324	6339	6353	6368	6383	6397	6412	6427	6442
.81	6457	6471	6486	6501	6516	6531	6546	6561	6577	6592
.82	6607	6622	6637	6653	6668	6683	6699	6714	6730	6745
.83	6761	6776	6792	6808	6823	6839	6855	6871	6887	6902
.84	6918	6934	6950	6966	6982	6998	7015	7031	7047	7063
.85	7079	7096	7112	7129	7145	7161	7178	7194	7211	7228
.86	7244	7261	7278	7295	7311	7328	7345	7362	7379	7396
.87	7413	7430	7447	7464	7482	7499	7516	7534	7551	7568
.88	7586	7603	7621	7638	7656	7674	7691	7709	7727	7745
.89	7762	7780	7798	7816	7834	7852	7870	7889	7907	7925
.90	7943	7962	7980	7998	8017	8035	8054	8072	8091	8110
.91	8128	8147	8166	8185	8204	8222	8241	8260	8279	8299
.92	8318	8337	8356	8375	8395	8414	8433	8453	8472	8492
.93	8511	8531	8551	8570	8590	8610	8630	8650	8670	8690
.94	8710	8730	8750	8770	8790	8810	8831	8851	8872	8892
.95	8913	8933	8954	8974	8995	9016	9036	9057	9078	9099
.96	9120	9141	9162	9183	9204	9226	9247	9268	9290	9311
.97	9333	9354	9376	9397	9419	9441	9462	9484	9506	9528
.98	9550	9572	9594	9616	9638	9661	9683	9705	9727	9750
.99	9772	9795	9817	9840	9863	9886	9908	9931	9954	9977
	0	1	2	3	4	5	6	7	8	9

APPENDIX IV

ANSWERS
TO PROBLEMS

Chapter 2 **CHEMICAL KINETICS**

1. In this reaction, the concentrations of A, B, X, and P would be expected to vary with time as illustrated in the accompanying figure.

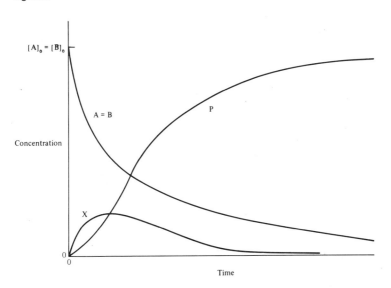

2. a. The observed *pseudo* first-order rate constant is $k_{obsd} = 9.7 \times 10^{-1}$ min^{-1}.

b. The half-life is $t_{1/2} = 0.71$ min.

3. a. The data indicate that the reaction is second-order.

b. The observed second-order rate constant is $k_{obsd} = 0.0445$ M^{-1} min^{-1}.

c. The half-life is $t_{1/2} = 585$ min.

4. a. The second-order rate constant is calculated from a plot of k_{obsd} versus phosphate buffer concentration. It is $k_B = 32$ M^{-1} sec^{-1}.

b. The observed *pseudo* first-order rate constants, k_{obsd}, are the sum of the rate constants for buffer-catalyzed reaction and for spontaneous reaction. If one divides the observed *pseudo* first-order rate constants by buffer concentration to obtain the second-order rate constant for buffer catalysis, the value obtained would be erroneously large since the spontaneous rate would not have been corrected for.

Chapter 3 THERMODYNAMICS OF CHEMICAL REACTIONS

1. The standard-state thermodynamic parameters of this ionization are: ΔH^0 = 9.0 kcal/mole, ΔG^0 = 9.8 kcal/mole, and ΔS^0 = -2.7 e.u.

2. a. The activation parameters are:$\Delta H^{\ddagger}$ = 27.0 kcal/mole, $\Delta G^{\ddagger}$ = 26.8 kcal/mole, $\Delta S^{\ddagger}$ = 0.7 eu.

b. The magnitude of the $\Delta S^{\ddagger}$ value for this reaction suggests that the rate limiting step is monomolecular. In reality, compound (I) undergoes a slow, rate-limiting rearrangement before a rapid reaction with water during its hydrolysis (D. Piszkiewicz and T. C. Bruice, *J. Am. Chem. Soc.* **90**, 2156 (1968)).

3. The activation parameters are: $\Delta H^{\ddagger}$ = 6.2 kcal/mole, $\Delta G^{\ddagger}$ = 20.6 kcal/mole, and $\Delta S^{\ddagger}$ = -48.3 e.u. Note that the large negative entropy of activation is in accord with this reaction being multimolecular.

Chapter 4 CATALYSIS IN AQUEOUS SOLUTION

1. Often there is more than one way to solve a problem such as a derivation. Here is one possible solution to this problem.
    In the text, it is demonstrated that

$$\frac{k_{obsd}}{[B]_{total}} = k_{BH} \left(\frac{[H^+]}{K_a + [H^+]} \right).$$ (41)

If

$$\frac{k_{obsd}}{[B]_{total}} = \frac{k_{BH}}{2},$$

then

$$\frac{k_{BH}}{2} = k_{BH} \left(\frac{[H^+]}{K_a + [H^+]} \right),$$

or

$$\frac{K_a + [H^+]}{2} = [H^+],$$

and at $k_{obsd} = k_{BH}/2$, $K_a = [H^+]$, or $pK_a = pH$.

2. At low pH

$$\frac{k_{obsd}}{[B]_{total}} = k_{BH},$$ (42)

and

$$\log \frac{k_{obsd}}{[B]_{total}} = \log k_{BH}.$$ (43)

At high pH

$$\frac{k_{obsd}}{[B]_{total}} = \frac{k_{BH}[H^+]}{K_a},$$ (44)

and

$$\log \frac{k_{obsd}}{[B]_{total}} = \log k_{BH} - \log K_a + \log [H^+].$$ (45)

By equating Eqs. (44) and (46) from the text,

$$\log k_{BH} = \log k_{BH} - \log K_a - pH,$$

and $-\log K_a = pH$, or $pH = pK_a$.

3. The constants involved are:

$\log k_H = 0.25$, $k_H = 1.78$ $M^{-1}days^{-1}$

$\log k_{H_2O} = -1.89$, $k_{H_2O} = 1.3 = 10^{-2}days^{-1}$

$\log k'_{H_2O} = -0.90$, $k'_{H_2O} = 1.2 \times 10^{-1}days^{-1}$

$\log k_{OH} = 3.90$, $k_{OH} = 7.95 \times 10^3$ M^{-1} $days^{-1}$.

4. a. Four possible rate laws are as follows:

$$\frac{k_{obsd}}{[B]_{total}} = k_1\left(\frac{[H^+]}{K_{app} + [H^+]}\right) + k_2\left(\frac{K_{app}}{K_{app} + [H^+]}\right)$$

$$\frac{k_{obsd}}{[B]_{total}} = k_1[H^+]\left(\frac{K_{app}}{K_{app} + [H^+]}\right) + k_2[OH^-]\left(\frac{[H^+]}{K_{app} + [H^+]}\right)$$

$$\frac{k_{obsd}}{[B]_{total}} = k_1\left(\frac{[H^+]}{K_{app} + [H^+]}\right) + k_2[OH^-]\left(\frac{[H^+]}{K_{app} + [H^+]}\right)$$

$$\frac{k_{obsd}}{[B]_{total}} = k_1[H^+]\left(\frac{K_{app}}{K_{app} + [H^+]}\right) + k_2\left(\frac{K_{app}}{K_{app} + [H^+]}\right).$$

b. The kinetically apparent pK_{app} is 6.8. This correlates well with the second ionization of phosphoric acid which has a pK_a of 6.7.

5.   The simplest rate law which fits the data is

$$k_{obsd} = k_H[H^+] + k_{-1}\left(\frac{[H^+]}{K_{app} + [H^+]}\right) + k_{-2}\left(\frac{K_{app}}{K_{app} + [H^+]}\right) + k_{OH}[OH^-],$$

where k_{-1} and k_{-2} are the first-order rate constants for the spontaneous hydrolysis of the mono- and di-ionized forms of acetyl phosphate. The constants involved are:

$\log k_H = 5.3$, $k_H = 2.0 \times 10^5$ M^{-1} min^{-1}

$\log k_{-1} = -4.0$, $k_{-1} = 10^4$ min^{-1}

$\log k_{-2} = -3.5$, $k_{-2} = 3.2 \times 10^3$ min^{-1}

$\log k_{OH} = 6.9$, $k_{OH} = 7.9 \times 10^6$ M^{-1} min^{-1}

$pK_{app} = 4.0$, $K_{app} = 10^{-4}$ M.

6.   a. The pH dependence for the hydrolysis of compound (I) may be described by

$$k_{obsd} = k_H[H^+]\left(\frac{[H^+]}{K_{app} + [H^+]}\right) + k_H[H^+]\left(\frac{K_{app}}{K_{app} + [H^+]}\right), \qquad (a)$$

or

$$k_{obsd} = K_H[H^+]\left(\frac{[H^+]}{K_{app} + [H^+]}\right) + k_0\left(\frac{[H^+]}{K_{app} + [H^+]}\right). \qquad (b)$$

   b. The first term of both equations which predominates at pH< 4 describes specific acid-catalyzed hydrolysis of compound (I) which has its carboxyl group in the un-ionized form:

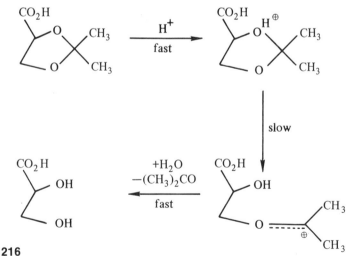

The second term of equation (a) which predominates at pH > 4 describes specific acid-catalyzed hydrolysis of compound (I) which has its carboxyl group in the ionized form. The mechanism of hydrolysis at pH > 4 is essentially identical with that which takes place at pH < 4 (above).

The second term of equation (b) which predominates at pH > 4 describes intramolecular general acid catalysis in the hydrolysis of compound (I).

7. a. The Brønsted β for this reaction is $\beta = 0.80$.
    b. The estimated rate constant is $k_2 = 2.25 \times 10^{-2}$ M^{-1} min^{-1}.

8. $\sigma_{meta} = + 0.75$ and $\sigma_{para} = +0.77$.

9. a. The value of ρ for formation of semicarbazones is $\rho = 0.91$.
    b. A positive value of ρ such as this indicates that the reaction is assisted by electron-withdrawing substituents. In this case, the carbonyl group, which is depleted of electrons, becomes more susceptible to nucleophilic attack by semicarbazide. Therefore, electron-withdrawing substituents accelerate the rate of semicarbazone formation.

Chapter 5 **ENZYME CATALYSIS AND KINETICS**

1. At $1/v = 0$ the intercept of the horizontal axis is $1/[S]$. If $1/v = 0$ the Lineweaver-Burk equation is

$$\frac{1}{v} = 0 = \frac{K_m}{V_{max}} \left(\frac{1}{[S]} \right) + \frac{1}{V_{max}},$$

or

$$\frac{K_m}{V_{max}} \left(\frac{1}{[S]} \right) = -\frac{1}{V_{max}},$$

and

$$\frac{1}{[S]} = -\frac{1}{V_{max}} \frac{V_{max}}{K_m} = -\frac{1}{K_m}.$$

2. a. The purpose of presenting this problem is to demonstrate that even the classical Michaelis-Menten derivation has a kinetically equivalent alternative. Since [S] is greater than $[E]_{total}$, free [S] is equal to the total [S]; and the velocity of this reaction is

$$v = k([E]_{total} - [ES])[S]. \tag{a}$$

If

$$K_d = \frac{([E]_{total} - [ES])[S]}{[ES]}, \tag{b}$$

then

$$([E]_{total} - [ES])[S] = K_d[ES]. \tag{c}$$

Substituting equation (c) into equation (a) yields

$$v = kK_d[ES]. \tag{d}$$

But we need [ES] in terms of $[E]_{total}$ and [S]. This may be obtained by expanding equation (b) to give

$$K_d[ES] = [E]_{total}[S] - [ES][S],$$

or

$$K_d[ES] + [ES][S] = [E]_{total}[S].$$

Factoring out [ES] yields

$$[ES](K_d + [S]) = [E]_{total}[S],$$

and

$$[ES] = \frac{[E]_{total}[S]}{K_d + [S]}. \tag{e}$$

Substituting equation (e) into equation (d) gives

$$v = kK_d \ \frac{[E]_{total}[S]}{K_d + [S]}.$$

This equation is similar in form to Eq. (10) in the text where

$$V_{max} = k_3[E]_{total} = kK_d[E]_{total}.$$

b. The equation describing the alternative mechanism to the Michaelis-Menten model reduces to

$$v = \frac{V_{max}[S]}{K_d + [S]}.$$

Therefore, at $v = V_{max}/2$, $[S] = K_d$.

3. a. The rate of ribonuclease-catalyzed formation of o-nitrophenylate from o-nitrophenyl oxalate indicates a two-phase, burst-turnover reaction. It suggests the formation of a covalent acyl-ribonuclease intermediate.

b. A possible simplified mechanism of this reaction is

Note that in the acylation step the negative charge of the carboxyl group of the substrate binds to the positive charge of histidine-12, thereby orienting the substrate at the active site.

4. For this reaction $K_m = 8 \times 10^{-4}$ M, $V_{max} = 1.25 \times 10^3$, $V_{max}/K_m = 1.56 \times 10^6$, and the inhibitor acts noncompetitively with a value of $K_I = 10^{-4}$ M.

5.   a. The inhibition of fumarase by succinate is competitive with a value of $K_1 = 5.2 \times 10^{-2}$ M.

b. The variation of V'_{max}/K'_m with pH describes a bell-shaped curve which suggests that catalysis is dependent upon two ionizable groups. A plot of log (V'_{max}/K'_m) versus pH suggests that in the uncomplexed enzyme these two groups are a general base with $pK_{e1} = 5.7$ and a general acid with $pK_{e2} = 7.1$.

c. The enthalpy of ionization of the group described by pK_{e1} suggests that it is a carboxyl group; the enthalpy of ionization of the group described by pK_{e2} suggests that it is the imidazolium side chain of a histidine residue.

d. One possible mechanism which is consistent with the data given is as follows:

Note that the data given in this problem are too fragmentary to allow any unambiguous solution. Many alternative solutions could be proposed. For example, the data presented here do not rule out the possibility that the group acting as general acid is the undissociated carboxyl, and the group acting as a general base is the neutral imidazole. However, the data do allow construction of the working hypothesis presented here which can be used as a basis for designing further experiments.

Chapter 6 **MULTIPLE SUBSTRATE REACTIONS**

1.   In the terminology of Cleland, the mechanism of catalysis by chymotrypsin is Ordered Uni Bi. It can be diagrammed as follows:

2.   a. The 47 simplest answers to this problem are:
Random Ter Ter. (This term describes only one unique mechanism; however, its diagrammatic representation is too complicated to make drawing it out worth the effort.)
Ordered Ter Ter. (There are six possible orders of substrate addition and six possible orders of substrate release giving a total of 36 different possible mechanisms.):

(amino acid and ATP and tRNA) (PP$_i$ and AMP and amino acyl-tRNA)

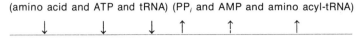

Ordered Bi Uni Uni Ordered Bi Ping Pong. (This mechanism has four possible permutations.):

(ATP and amino acid) PP$_i$ tRNA (AMP and amino acyl-tRNA)

↓ ↓ ↑ ↓ ↑ ↑ .

Uni Uni Ordered Bi Ordered Bi Ping Pong. (This mechanism has four possible permutations.):

ATP PP$_i$ (amino acid and tRNA) (AMP and amino acyl-tRNA)

↓ ↑ ↓ ↓ ↑ ↑ .

Random Bi Uni Uni Random Bi Ping Pong:

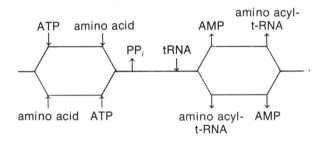

Uni Uni Random Bi Random Bi Ping Pong:

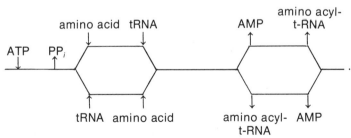

b. According to the description given, the mechanism of the amino acyl-tRNA synthetase is one of the five possible Bi Uni Uni Bi Ping Pong mechanisms.

221

3. a. The pattern of this double reciprocal plot is consistent with the enzyme, ATP, and isoleucine forming a ternary complex in the reaction mechanisms.

b. Assuming the formation of an enzyme-ATP-amino acid ternary complex, these data suggest that tRNA interacts with the enzyme in a Ping Pong relationship during the course of the enzyme-catalyzed reaction. When considered together, the data presented in parts a and b of this problem are in accord with the Bi Uni Uni Bi Ping Pong mechanism, which is generally accepted for the amino acyl-tRNA synthetases.

4. The product inhibition patterns for the Ping Pong Bi Bi mechanism are as follows:

Product	Vary A		Vary B	
inhibitor	Unsat	Sat B	Unsat	Sat A
P	NC	—	Comp	Comp
Q	Comp	Comp	NC	—

5. a. These double reciprocal plots are consistent with a Sequential mechanism.

b. The product inhibition data are consistent with an Ordered Sequential mechanism in which NADPH binds to the enzyme before α-acetolactate, and in which α,β-dihydroxyisovalerate (DHIV) is released before NADP:

$$\text{NADPH} \quad \alpha\text{-acetolactate} \quad \text{DHIV} \quad \text{NADP}$$
$$\downarrow \qquad\qquad \downarrow \qquad\qquad \uparrow \qquad\qquad \uparrow$$

6. The secondary plot of Lineweaver-Burk plot Slopes follows the equation

$$\text{Slope} = \frac{K_m^A}{V'} + \frac{K_m^A[\text{I}]}{V'K_{\text{I slope}}}.$$

At slope = 0,

$$0 = \frac{K_m^A}{V'} + \frac{K_m^A[\text{I}]}{V'K_{\text{I slope}}},$$

and

$$-\frac{K_m^A}{V'} = \frac{K_m^A[\text{I}]}{V'K_{\text{I slope}}}.$$

Dividing by K_m^A/V',

$$-1 = \frac{[I]}{K_{I\ slope}},$$

and

$$-K_{I\ slope} = [I].$$

Similarly, the secondary plot of Lineweaver-Burk plot Intercepts follows the equation

$$\text{Intercept} = \frac{1}{V'} + \frac{[I]}{V'K_{I\ intercept}}.$$

At Intercept $= 0$,

$$0 = \frac{1}{V'} + \frac{[I]}{V'K_{I\ intercept}},$$

and

$$-\frac{1}{V'} = \frac{[I]}{V'K_{I\ intercept}},$$

or

$$-K_{I\ intercept} = [I].$$

Chapter 7   METABOLIC REGULATION BY ENZYMES

1. a. The shape of the curve obtained in the absence of modifiers is sigmoid in shape, and it reflects a positive homotropic interaction.

b.	fructose-1,6-diphosphate	K-type activator
	serine	K-type activator
	alanine	K-type inhibitor
	phenylalanine	K-type inhibitor
	tryptophan	V-type inhibitor
	ATP	V-type inhibitor

2. a. The Lineweaver-Burk plot is concave upward.

   b. The Hill coefficient, n, is 1.85; $K' = 5.1 \times 10^{-8}$.

   c. The plot of v versus substrate concentration, the Lineweaver-Burk plot, and the Hill plot are consistent with the MWC, the KNF, the Frieden, and the Bernhard equilibrium models.

3. a. The Lineweaver-Burk plot is linear.

b. The Hill coefficient, n, is 1; and $K' = 1.5 \times 10^{-5}$.

c. The data presented in this problem indicate that in the presence fructose-1,6-diphosphate, there is no interaction between catalytic sites; the enzyme follows simple Michaelis-Menten kinetics.

4. a. Lineweaver-Burk plot is concave downward.

b. The Hill coefficient, n, is 0.55; $K' = 1.8 \times 10^{-2}$.

c. The shape of the Lineweaver-Burk plot and the slope of the Hill plot indicate a negative homotropic interaction or negative cooperativity in the presence of phenylalanine. These data are explained best by the KNF-sequential and the Bernhard equilibrium models.

5. a. If there is a single substrate, only the kinetic model described by Rabin is consistent with the data.

b. If there are two or more substrates, either of the two kinetic models, those of Rabin and Ferdinand, may explain the observed results.

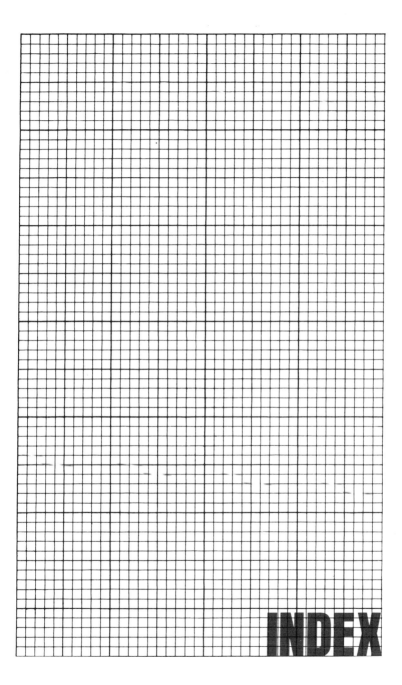

INDEX